Comprehensive Probability Actuarial Exams:
A Manual for SOA Exam P

First Edition

Written by: Digital Actuarial Resources, LLC

Copyright © 2010, by Digital Actuarial Resources, LLC

Publication Date: August 20, 2010
Format: Perfect-bound

Published by Digital Actuarial Resources, LLC
Saint Paul, Minnesota
www.digitalactuarialresources.com

All rights reserved. No part of this book may be reproduced in any form or by any means, mechanical or electronic, including but not limited to recording, photocopying, or using any information storage and retrieval system, without written permission from the publisher.

Printed in the United States of America

ISBN 978-1-4537807-9-4

Table of Contents

CHAPTER 1: INTRODUCTION TO PROBABILITY — 3

- SECTION 1-1: THE STUDY OF UNCERTAINTY — 3
- SECTION 1-2: SET THEORY — 5
- SECTION 1-3: VENN DIAGRAMS — 15
- SECTION 1-4: INDEPENDENT EVENTS — 18
- SECTION 1-5: CONDITIONAL PROBABILITY — 21
- SECTION 1-6: LAW OF TOTAL PROBABILITY AND BAYES' THEOREM — 28

CHAPTER 2: COUNTING TOOLS — 33

- SECTION 2-1: COUNTING BY MULTIPLICATION — 34
- SECTION 2-2: PERMUTATIONS AND COMBINATIONS — 35
- SECTION 2-3: MULTINOMIAL COMBINATIONS — 40
- SECTION 2-4: DISCRETE UNIFORM DISTRIBUTION — 42

CHAPTER 3: BASICS OF PROBABILITY DISTRIBUTIONS — 44

- SECTION 3-1: RANDOM VARIABLES — 44
- SECTION 3-2: STRUCTURE OF DISCRETE PROBABILITY DISTRIBUTIONS — 46
- SECTION 3-3: STRUCTURE OF CONTINUOUS PROBABILITY DISTRIBUTIONS — 47
- SECTION 3-4: DISTRIBUTION FUNCTIONS — 51
- SECTION 3-5: THE QUANTILE FUNCTION — 59
- SECTION 3-6: BERNOULLI DISTRIBUTION — 63
- SECTION 3-7: BINOMIAL DISTRIBUTION — 66
- SECTION 3-8: CONTINUOUS UNIFORM DISTRIBUTION — 70

CHAPTER 4: MULTIVARIATE DISTRIBUTIONS — 73

- SECTION 4-1: BASICS OF BIVARIATE DISTRIBUTIONS — 73
- SECTION 4-2: COMPUTING MARGINAL DISTRIBUTIONS — 81
- SECTION 4-3: DISTRIBUTIONS WITH MORE THAN TWO RANDOM VARIABLES — 86
- SECTION 4-4: DETERMINING THE DISTRIBUTION FOR Y DEPENDING ON X — 87
- SECTION 4-5: DETERMINING THE DISTRIBUTION FOR MANY Y'S DEPENDING ON MANY X'S — 92
- SECTION 4-6: CONDITIONAL PROBABILITY DISTRIBUTIONS — 104

CHAPTER 5: ANALYZING DISTRIBUTIONS — 109

- SECTION 5-1: EXPECTED VALUE — 109
- SECTION 5-2: CONDITIONAL EXPECTED VALUE — 119
- SECTION 5-3: MEDIAN — 124

SECTION 5-4: MODE	129
SECTION 5-5: VARIANCE	131
SECTION 5-6: MEASURES OF LINEAR RELATIONSHIP—COVARIANCE AND CORRELATION	135
SECTION 5-7: MOMENT-GENERATING FUNCTIONS	141

CHAPTER 6: COMMON DISCRETE DISTRIBUTIONS — 148

SECTION 6-1: BERNOULLI AND BINOMIAL DISTRIBUTIONS REVISITED	148
SECTION 6-2: POISSON DISTRIBUTION	153
SECTION 6-3: NEGATIVE BINOMIAL AND GEOMETRIC DISTRIBUTIONS	158
SECTION 6-4: HYPERGEOMETRIC DISTRIBUTION	166
SECTION 6-5: MULTINOMIAL DISTRIBUTION	170

CHAPTER 7: COMMON CONTINUOUS DISTRIBUTIONS — 175

SECTION 7-1: CONTINUOUS UNIFORM DISTRIBUTION REVISITED	175
SECTION 7-2: NORMAL DISTRIBUTION	178
SECTION 7-3: CORRECTION FOR CONTINUITY	184
SECTION 7-4: CENTRAL LIMIT THEOREM	187
SECTION 7-5: BIVARIATE NORMAL DISTRIBUTION	193
SECTION 7-6: GAMMA DISTRIBUTION	198
SECTION 7-7: BETA DISTRIBUTION	207
SECTION 7-8: CONJUGATE PRIORS	211

APPENDIX A: STANDARD NORMAL DISTRIBUTION — 222

APPENDIX B: COMMON SERIES — 223

Chapter 1: Introduction to Probability

This chapter establishes the basic theorems and notation which are heavily used in probability analysis. The first chapter primarily covers the probability of events; the probability for random variables is reserved for later chapters. Most of the theorems for events can be extended to random variables. There are three axioms for probability that must hold for all random variables and events. The chapter gives an in-depth review of sets and Venn diagrams. It also gives the distinction between independent and dependent events. The chapter ends with an examination of Bayes' Theorem.

Section 1-1: The Study of Uncertainty

Probability is a branch of math that provides tools to analytically examine the chances of events occurring. Probability studies uncertainty in life. It helps us make decisions. Likewise, observations can aid us in making inferences about new people or things. Probability utilizes a large collection of math, from basic algebra and geometry to differential equations, topology, and combinatorial analysis. Probability overlaps significantly with statistics and is often considered the base for statistics. The study of probability is highly intertwined. All the early topics contribute to later topics.

Additionally, sometimes later ideas help to explain early concepts. Probability began as a branch of math in the seventeenth century. However, most theorems and terminology in this book date to the early 1900's at the earliest.

Probabilities translate to numbers. As a real number, a probability must be between 0.0 and 1.0, inclusive. If an event cannot happen, its probability is zero. If an event always happens, its probability is one. As corresponding percentages, probabilities remain in the range 0% to 100%. The abbreviated notation for probability is "Pr" or just 'P.' For instance, the probability of event A is represented as Pr(A).

N(A) = the number of elements in set A
Pr(A) = the probability that an outcome within set A happens

A distinction exists between a population and a sample. A *population* is the entire group of objects under examination. For example, the population might be all 25,000 undergraduate students at a university. The population may also be termed the "study region." A *sample* is a subset of the population. For instance, a sample might be 50 chosen undergraduate students. The researcher typically chooses members of the sample randomly. In a random selection, each member of the population has an equal chance of being chosen. Sampling saves money and time. The population could also be infinite in size, leading to difficulties sampling each unit. Certain members of the population could be unreachable.

A difference is also present between a statistic and a global parameter. *Statistics* really apply to just the sample. The population has *global parameters* which are typically unknown precisely. The sample statistics estimate the global parameters. For example, a sample statistic might be the sample mean weight of a person from a small group of people. The sample mean approximates the true average weight in the bigger population. The population is usually too big to estimate its parameters precisely. Even if a researcher could gather data on every member of the population, the population is typically dynamic. The population changes in size. The members themselves move. If the population is a group of people, their views and attitudes are continually shifting. Therefore, in theory, it is impossible to precisely know the global parameters.

Experiments

An *experiment* is a happening that produces an observable outcome. The experiment could be planned or unplanned. The experiment might simply be the observation of some phenomenon. For instance, an experiment could be a well-controlled scientific test, a sociology survey, or observations of nature. The experiment produces precisely one outcome per run. However, the outcome does not need to be a single value; the outcome could be an ordered tuple or a set of attributes for many variables.

One execution of an experiment could have many internal trial runs. The results of the trial runs will be aggregated to produce the experiment's outcome. Consider a basketball player throwing free throws. The experiment could contain 20 free throw attempts (or trial runs). Each free throw could miss or score. The outcome of the experiment could be the proportion of shots made.

In general, as you increase the amount of trials within an experiment, the results become more accurate. More trials lead to stronger estimates of the true population parameters and probabilities. If you are trying to estimate the probability of event E, you will develop an answer very close to the real probability as the amount of trials becomes sufficiently big.

The universe of all possible outcomes for an experiment is called the *sample space*. It contains every potential element in the realm of the problem. Objects in the sample space can be called points, sample points, outcomes, or elementary outcomes. The sample space could hold infinitely many objects or a countable quantity. The letter 'S' symbolizes the sample space. The sample space is a set. Each different type of experiment will entail a different sample space. For example, the sample space could be a small list of colors, the entire real line, or a collection of integers. When defining a sample space, you could explicitly list every element in S. For instance, if you flip a coin three times and are concerned about the sequence of heads and tails, you could provide the sample space as {(H, H, H), (H, H, T), (H, T, H), …, (T, T, T)}. However, when the sample space is very large, it is unwise to list every element. An implicit rule could help define the sample space.

An *event* is a subset of outcomes from the sample space. Events are typically denoted with capital letters, such as A, B, and C. An event could hold a single outcome, a few outcomes, or it could equal the entire sample space. An event with just one potential outcome is called a *simple event*. If the experiment yields one of the outcomes in event A, then event A has occurred. It is possible for multiple events to occur in the same run of an experiment, assuming all the events share the same outcome. For any run of an experiment, the sample space always happens as an event (i.e., Pr(S) = 1).

Section 1-2: Set Theory

A *set* is a collection of elements. Mathematicians use curly braces, { }, to enclose the objects in the set. Example sets include {2, 4, 6, 8, 10} and {Bemidji, St. Paul, Duluth, International Falls}. Sets typically contain numbers, but they can also hold attributes or qualitative values. Sets are often represented with capital letters, such as A and B. Elements are in lowercase, often with a subscript to denote the element's index. For example, the sixth element in set A is a_6. The notation matches that of sequences.

Sets are classified according to how many and how densely packed the elements are. The three primary types of sets are finite sets, countably infinite sets, and uncountably infinite sets. A *finite set* contains a quantity of values which a person can specify with a positive integer below infinity. For example, a set with 50 elements is a finite set. A *countably infinite* set has infinitely many elements, but each element can be paired in a one-to-one fashion with the set of positive integers. Examples of countably infinite sets include the set of positive integers itself, the set of rational numbers, and any sequence. Finally, an *uncountably infinite* set has an infinite amount of values, and each value cannot be paired in a one-to-one fashion with the positive integers. A prime example of an uncountably infinite set is any continuous interval (open or closed), such as [0, 1]. For an uncountably infinite set, given two arbitrarily close points within an interval, you can find infinitely many additional points between the two chosen points. No matter how small you attempt to partition a continuous interval, the resulting sub-interval will still have infinitely many points.

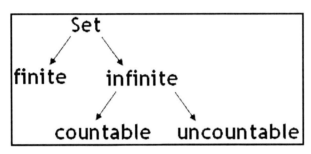

The *empty set* is represented with void parentheses or the symbol $\emptyset$. This set contains zero outcomes. It is also called the *null set*. A simple fact about the null set is that $\Pr(\emptyset) = 0$. In other words, it is impossible for an experiment to not produce an outcome.

The *union* of two sets A and B is a new set containing all elements in A, in B, or in both A and B. The union operation consolidates two sets. The union operator is $\cup$, like the letter 'U.' The union operation does not duplicate elements. That is, if A and B both contain '5,' then only one copy of '5' will be in the union set. You can view the union operation as a function that outputs a new set. Union corresponds to "OR." It finds the elements that belong in one set or the other or both sets.

The *intersection* of two sets A and B is a new set containing all elements that exist in both A and B. The intersection operator is $\cap$. The intersection operator also does not duplicate outcomes. Intersection corresponds to "AND." The operator finds elements that belong in the first set and the second set. An abbreviated form of writing $A \cap B$ is AB.

Example
(1.) Find the union of {8, 23, 60, 130, 2} and {60, 5, 82, 130}
(2.) Find the intersection of {6, 20, 9, 34, 90, 3, 1} and {92, 20, 45, 3, 29, 7, 6}

> Solutions:
> (1.) Copy the entire first set and augment it with new elements that appear in the set on the right side.
> {8, 23, 60, 130, 2, 5, 82}
>
> (2.) Which elements exist in both sets? {6, 20, 3}

> **Example**
> Evaluate these expressions:
> (1.) $\{13, 11, 14, 9, 6\} \cup \{7, 14, 4, 15, 11\}$
> (2.) $\{38, 32, 22, 8, 5, 20\} \cap \{36, 29, 5, 38, 40\}$
>
> Solutions:
> (1.) {13, 11, 14, 9, 6, 7, 4, 15}
> (2.) {38, 5}

Two sets, A and B, are *disjoint* if they have no elements in common. Another term for disjoint is *mutually exclusive*. Disjoint sets have absolutely no overlap. In other words, sets A and B are disjoint iff $A \cap B = \emptyset$.

Disjointedness can apply to a larger collection of sets. A collection of n sets, where n > 2, are *pairwise disjoint* when none of the sets have any elements in common. That is, you can compare any two sets from the collection and find that they have no overlap. In mathematical terms, events $E_1, E_2, \ldots, E_n$ are *pairwise disjoint* iff E_i and E_j are mutually exclusive ($\forall \; i \neq j$).

> **Example**
> Are these sets disjoint?
> (a.) {18, 4, 15, 9, 15}, {12, 10, 20, 4, 16}
> (b.) {-9, 13, 11, -10, -7}, {-4, -4, 12, -1, 0}
> (c.) {32, 31, 28, 39}, {37, 36, 31, 33, 32}
> (d.) {1, 7, 15}, {3, 4, 8}, {2, 19}, {6, 24, 0, 11}
>
> Solutions:
> (a.) No; 4 appears in both   (b.) Yes   (c.) No; 31 & 32 appear in both   (d.) Yes

When two events A and B are mutually exclusive, the probability that A or B happens is simply the sum of the individual probabilities for the events. We do not need to be concerned about the chance that A and B both happen. Similarly, if we want to compute the number of elements in sets A and B, we can just add the number of elements in each set.

Adding-Counting Principle
Suppose that sets A and B are mutually exclusive. Then,
- (1.) $N(A \cup B) = N(A) + N(B)$
- (2.) $\Pr(A \cup B) = \Pr(A) + \Pr(B)$

The formula below extends the basic adding-counting principle to n mutually exclusive events:

Adding Principle for Pairwise Mutual Exclusion
Suppose that events $E_1, E_2, \ldots, E_n$ are pairwise mutually exclusive. Then,
$$N(E_1 \cup E_2 \cup \ldots \cup E_n) = N(E_1) + N(E_2) + \ldots + N(E_n)$$
$$\Pr(E_1 \cup E_2 \cup \ldots \cup E_n) = \Pr(E_1) + \Pr(E_2) + \ldots + \Pr(E_n)$$

Example
Suppose Enoch rolls two dice. What is the chance that the dice sum to 9 or 4?

Solution:
The outcome of the "experiment" is the sum between the two dice. It is impossible for dice to sum to 9 and to 4 in the same roll, so the events $sum_1 = 9$ and $sum_2 = 4$ are mutually exclusive. Note that the events are not complementary since $sum = 2, 3, 5$ etc. are also possibilities in the sample space.

You should list the possible values for the dice that result in sums of 9 or 4.
When sum = 9, the possible dice values are {3, 6}, {6, 3}, {4, 5}, {5, 4}.
When sum = 4, the possible dice values are {1, 3}, {3, 1}, {2, 2}

We can use the adding-counting principle because the events $sum_1 = 9$ and $sum_2 = 4$ are mutually exclusive. First, find the probability of each event (or sum). The sample space for two dice contains 36 potential configurations. Each configuration has the same chance of happening, so the probability of each configuration is 1/36.

$P(sum_1 = 9) = (1/36) * 4 = 4/36$
$P(sum_2 = 4) = (1/36) * 3 = 3/36$

$P((sum_1 = 9) \cup (sum_2 = 4)) = P(sum_1 = 9) + P(sum_2 = 4)$
$= 4/36 + 3/36$
$= 7/36 \approx 0.1944 \approx 19.44\%$

When sets A and B are not mutually exclusive, and we are computing the probability of the union of A and B, we need to make an adjustment for the overlap. The adjustment consists of subtracting the probability that both A and B happen. We must subtract the probability of the intersection because $\Pr(A \cap B)$ is counted twice in $\Pr(A) + \Pr(B)$.

Adding-Counting Principle under Non-Mutual Exclusion
Suppose that sets A and B are not mutually exclusive. Then,
(1.) $N(A \cup B) = N(A) + N(B) - N(A \cap B)$
(2.) $\Pr(A \cup B) = \Pr(A) + \Pr(B) - \Pr(A \cap B)$

The *complement* of set A is all elements that lie outside set A (and within S). The complement of A is denoted by $\overline{A}$ (pronounced "not A"). Other ways to write the complement include A^C and ~A. The complement is essentially the NOT operator in logic. The complement operator "inverts" a set and returns all elements from the sample space which are not in A. All elements of S either lie within event A or within A's complement. A simple probability rule is:

Complementary Probability Theorem
$\Pr(A) + \Pr(A^C) = 1$

Example
Let S = {3, 6, 9, 12, 15, 18, 21, 24, 27, 30}.
Find the complement of each set below (the sets lie within S):
(a.) A = {12, 15, 18}
(b.) B = {30, 21, 9, 24}

Solutions:
(a.) A^C = {3, 6, 9, 21, 24, 27, 30}
(b.) B^C = {3, 6, 12, 15, 18, 27}

If the sample space contains only two events, and the events are mutually exclusive, then the two events are *complementary events*. Two events that are complementary are usually opposites of each other. Complementary events might be success versus failure, employed or unemployed, male versus female, and so on. Complementary events usually describe a common trait, like gender. Suppose that sets A and B are complementary. Then, $P(\overline{A}) = P(B)$ and $P(\overline{B}) = P(A)$.

Set A is a *subset* of set B when all the elements from A also reside in B. In addition, B will be a *superset* of A. Textbooks vary in their subset notation. However, we will assume that $A \subset B$ means that "A is a proper subset of B" (i.e., A will have a limited collection of elements from B, and A could never equal B). On the other hand, the notation $A \subseteq B$ reads "A is a subset of B," with the possibility that A = B. You can view the extra bar on the bottom as the possibility that A and B are identical.

Sets A_1, A_2, ... A_n form a *partition* of the sample space if they are all pairwise disjoint and they completely cover S. That is, $\Pr(A_1 \cup A_2 \cup ... \cup A_n) = \Pr(S) = 1$, and $(A_i \cap A_j) = \emptyset \;\; \forall \; i \neq j$. You can view the sample space broken into pieces without any overlap:

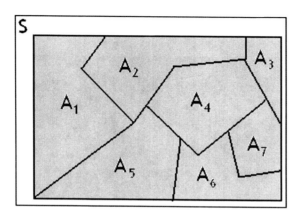

The distributive laws for sets aid in expanding expressions. In the equations below, the external operator next to A is applied to both B and C, and the two sub expressions are combined with the internal operator (between B and C).

Distributive Laws
(1.) $A \cap (B \cup C) = (A \cap B) \cup (A \cap C)$
(2.) $A \cup (B \cap C) = (A \cup B) \cap (A \cup C)$

DeMorgan's Laws help to simplify a union or intersection expression that is complemented. If the expression $(A \cap B)$ is false, then A is false, B is false, or both A and B are false. Likewise, if the expression $(A \cup B)$ is false, then neither A nor B can be true.

DeMorgan's Laws
(1.) $\overline{A \cap B} = \overline{A} \cup \overline{B}$
(2.) $\overline{A \cup B} = \overline{A} \cap \overline{B}$

DeMorgan's Laws can be expanded to cover more than two events. For instance, the expression $\overline{(A \cup B \cup C)}$ means that none of A, B, and C occurred. You can translate the expression as $\overline{A} \cap \overline{B} \cap \overline{C}$.

Probability of the Union of 3 Events
Let A, B, and C be events, which may/may not be pairwise disjoint. Then,

$$\Pr(A \cup B \cup C) = \Pr(A) + \Pr(B) + \Pr(C) - \Pr(AB) - \Pr(AC) - \Pr(BC) + \Pr(ABC)$$

The diagram below shows the progression of the formula for finding the probability of the union for 3 events. The number in each region (e.g., +2) shows the number of times that the region is counted after each stage. The objective is to count each region exactly once, repairing overlap.

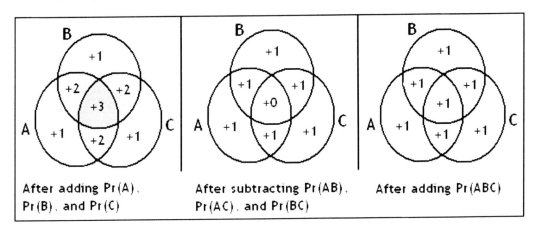

The Inclusion-Exclusion Theorem is a generalization for the probability of the union of n events. To compute the probability, first sum together the probability that each individual event happens. Then, subtract the probabilities for any two distinct events happening simultaneously. Then, add back the probabilities for any three distinct events occurring. This process of adding in some probability, shaving off a little probability, and adding back a bit of probability continues until the final probability term which describes the chance that all n events happen. If n is even, then the final term is subtracted from the computation. If n is odd, then the final term is added to the total. When all the events are pairwise disjoint, you can disregard all the terms with 2 or more events because the probability of the intersection of any collection of events will be 0.

Inclusion-Exclusion Theorem
Let E1, E2, ..., En be n events which may/may not be pairwise disjoint. Then,

$$\Pr(E_1 \cup E_2 \cup ... \cup E_n) = \sum_{i=1}^{n} \Pr(E_i) - \sum_{i \neq j} \sum \Pr(E_i \cap E_j) + \sum_{i \neq j \neq k} \sum \sum \Pr(E_i \cap E_j \cap E_k) - ...$$
$$..... \pm \Pr(E_1 E_2 ... E_n)$$

Example
Dredge rolls two dice. What is the chance that at least one die is over 4 or the dice sum to an even number?

Solution:
First, label each event.
Let A = at least one die (possibly both dice) is over 4
Let B = the dice sum to an even number

Events A and B are not mutually exclusive, since both events could happen at the same time.

Dice values that produce event A:
  {5, 1-6}, {6, 1-6}, {1-4, 5}, {1-4, 6}   (20 total configurations)

	die 2					
die 1	1	2	3	4	5	6
1	2	3	4	5	6	7
2	3	4	5	6	7	8
3	4	5	6	7	8	9
4	5	6	7	8	9	10
5	6	7	8	9	10	11
6	7	8	9	10	11	12

Dice values that produce event B:
  {1, 1}, {1, 3}, {1, 5}, {2, 2}, {2, 4}, {2, 6}, {3, 1}, {3, 3}, {3, 5}, {4, 2}, {4, 4},
  {4, 6}, {5, 1}, {5, 3}, {5, 5}, {6, 2}, {6, 4}, {6, 6}   (18 total configurations)

	die 2					
die 1	1	2	3	4	5	6
1	2	3	4	5	6	7
2	3	4	5	6	7	8
3	4	5	6	7	8	9
4	5	6	7	8	9	10
5	6	7	8	9	10	11
6	7	8	9	10	11	12

Dice values that produce both events A and B:
  {1, 5}, {2, 6}, {3, 5}, {4, 6}, {5, 1}, {5, 3}, {5, 5}, {6, 2}, {6, 4}, {6, 6}
  (10 total configurations)

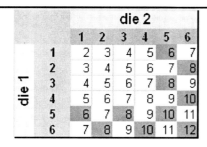

$P(A \cup B) = P(A) + P(B) - P(A \cap B)$
$= (20/36) + (18/36) - (10/36)$
$= 28 / 36 = 0.77778 = 77.778\%$

Example
Suppose you spin a wheel containing the values 1 through 20 in equal increments. Let the resulting value on the wheel be X and several events be:
  E_1: $X = 15$
  E_2: $3 \leq X < 9$
  E_3: $X \geq 16$

Evaluate these probabilities:
(a.) $P(E_1)$ (b.) $P(E_1 \cap E_2)$ (c.) $P(E_1 \cup E_2)$ (d.) $P(E_3)$

Solutions:
(a.) Each value on the wheel has the same chance of occurring. The probability of a particular value is $1/20 = 0.05$.
$P(E_1) = P(X = 15) = 0.05$

(b.) Events E_1 and E_2 are mutually exclusive; they can never occur together. Therefore, the chance of both events happening is 0.

(c.) $P(E_1 \cup E_2) = P(E_1) + P(E_2) = 0.05 + 6 * 0.05 = 0.35 = 35\%$.

(d.) $P(E_3) = P(X = 16 \text{ OR } X = 17 \text{ OR } X = 18 \text{ OR } X = 19 \text{ OR } X = 20) = 5 * 0.05 = 0.25$

Example
Ophir is running a unique game of chance. It uses a dial/wheel with three equally sized regions labeled red, green, and blue. Ophir also tosses a fair die with four faces labeled A, B, C, and D.

(a.) Write the sample space for the experiment.
(b.) What is the probability of the outcome (blue, C)?
(c.) What is the chance of obtaining "red" on the dial?
(d.) What is the probability that the die is 'A' or 'D?'

Solutions:

(a.) {(red, A), (red, B), (red, C), (red, D), (green, A), (green, B), (green, C), (green, D), (blue, A), (blue, B), (blue, C), (blue, D)}
(b.) The sample space contains 12 equally likely outcomes. Therefore, the probability of an individual outcome is 1/12 = 8.33%.
(c.) The events of spinning the dial and tossing the die are independent. Therefore, the chance of obtaining a certain colored region on the dial is independent of the value on the die. You can work with the outcome of the dial completely separately from the die. The dial has three equally sized regions, so the chance of each region is 1/3. Therefore, the chance of "red" is 1/3.
(d.) Again, the dial and die are independent. The die has four equally likely faces. Each face has probability 1/4 of occurring. Therefore, the probability of side 'A' or side 'D' is $2*(1/4) = 0.5$.

Example
Dredge is conducting a sociology study on sleep patterns. He developed a few possible options for his question "During which interval of the day do you sleep most?":
A = Between 6 PM and midnight
B = Between midnight and 6 AM
C = Between 6 AM and 6 PM
D = I don't require sleep; this question doesn't apply to me

Dredge uncovered the following probabilities: P(A) = 0.35, P(B) = 0.58, P(C) = 0.04, P(D) = 0.03.

(a.) What is $P(\overline{B})$?   (b.) What is P(A OR C)?   (c.) What is $P(\overline{D}$ OR B)?
(d.) What is the chance the person sleeps primarily during a six-hour interval?
(e.) What is the chance the person needs some sleep?

Solutions:
All the events are mutually exclusive, since none of the hours overlap and the subject can only give one answer.
(a.) $P(\overline{B}) = 1 - P(B) = 1 - 0.58 = 0.42$
(b.) $P(A \cup C) = P(A) + P(C) = 0.35 + 0.04 = 0.39$
(c.)
$$P(\overline{D} \cup B) = P(\overline{D}) + P(B) - P(\overline{D} \cap B)$$
$$= (1 - P(D)) + P(B) - P(\overline{D}) * P(B)$$
$$= (1 - 0.03) + 0.58 - (1 - 0.03) * 0.58$$
$$= 0.97 + 0.58 - 0.97 * 0.58$$
$$= 1.55 - 0.5626$$
$$= 0.9874$$

(d.) $P(A \cup B) = P(A) + P(B) = 0.35 + 0.58 = 0.93$

(e.) $P(\overline{D}) = 0.97$

Basic Axioms of Probability

A few rules form the core of the study of probability. The first rule states that the probability of any event cannot be negative. The second rule forces the probability of an event to be less than or equal to 1.

Probability Axioms
(1.) $\Pr(A) \geq 0 \quad \forall$ events A.
(2.) $\Pr(S) = 1$.
(3.) For a set of pairwise disjoint events A1, A2, ..., An, $\Pr\left(\bigcup_{i=1}^{\infty} A_i\right) = \sum_{i=1}^{\infty} \Pr(A_i)$.

The third rule also holds for n pairwise disjoint events (rather than infinitely many):

$$\Pr\left(\bigcup_{i=1}^{n} A_i\right) = \sum_{i=1}^{n} \Pr(A_i)$$

Section 1-3: Venn Diagrams

Venn diagrams are used to graph events and their probabilities. The typical Venn diagram has two or three events. Including more than three events makes the graph hard to understand. A Venn diagram consists of a few basic elements. First, the entire sample space (S) is drawn as the background rectangle. Each event is then drawn as a circle. Some diagrams use smaller circles for less probable events, but you should not assume that circle size is related to the event's probability. Numerical probabilities might be written inside each region. When circles A and B overlap, events A and B can occur simultaneously (i.e., the outcome $(A \cap B)$ is possible). Likewise, if three circles overlap in a small region, then all three events could happen at the same time. The shown events may not cover all possible outcomes, so some unlisted outcome in the background could also happen.

Shown below is an example of a Venn diagram. It contains events A and B, graphed as circles. The region inside 'A' represents the probability of event A happening. The graph includes a few listed probabilities. The lightly shaded area on the left gives the probability of only A happening and not B. The shaded region in the center shows the chance that both events A and B occur. Finally, the graph accounts for the chance that neither A nor B happens, which is represented by the white background.

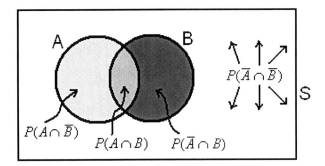

Regions that do not overlap are disjoint. For example, the outcomes $(A \cap \overline{B})$ and $(\overline{A} \cap B)$ are mutually exclusive and cannot happen at the same time. In fact, the four probabilities highlighted in the diagram are all disjoint. Remember that all the probabilities in a Venn diagram must sum to 1.0.

Venn diagrams are not essential to solving a problem. They primarily help you to visualize the situation. Once you become skilled in probability, you should work problems without drawing Venn diagrams. Drawing them increases the time required to solve a problem.

Example
Suppose an experiment has two events—A and B. These events hold all possible outcomes from the experiment. You know that the probability of only B happening is 0.37. The probability of only A happening is 0.49. Set up a Venn diagram and find the probability of A AND B happening. In addition, compute P(A) and P(B).

Solution:
The Venn diagram appears as:

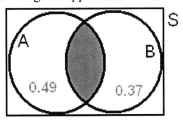

We need to find the probability in the shaded area, which represents $A \cap B$. Since A and B are the only two events in the sample space, and they are not mutually exclusive,
$P(S) = P(A \cap \overline{B}) + P(\overline{A} \cap B) + P(A \cap B)$
$1.0 = 0.49 + 0.37 + P(A \cap B)$
$P(A \cap B) = 0.14$

$P(A) = P(A \cap \overline{B}) + P(A \cap B) = 0.49 + 0.14 = 0.63$

$P(B) = P(\overline{A} \cap B) + P(A \cap B) = 0.37 + 0.14 = 0.51$

As a check, we should make sure the adding-counting principle under non-mutual exclusion holds:
$P(S) = 1.0 = P(A \cup B) = P(A) + P(B) - P(A \cap B)$
$\downarrow$
$1.0 = 0.63 + 0.51 - 0.14$
$1.0 = 1.0$
Checks

Example
A sample space has two events—A and B—among other possible events. It is known that $P(A) = 0.46$, $P(A \cap B) = 0.22$, and $P(\overline{A \cup B}) = 0.13$. Find $P(B)$ and $P(B \cap \overline{A})$.

Solution:
The initial Venn diagram appears as:

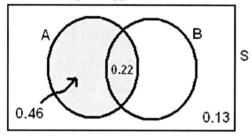

We can first find $P(B \cap \overline{A})$:
$P(S) = P(A) + P(B \cap \overline{A}) + P(\overline{A} \cap \overline{B})$
$1.0 = 0.46 + P(B \cap \overline{A}) + 0.13$
$P(B \cap \overline{A}) = 0.41$

$P(B) = P(A \cap B) + P(B \cap \overline{A}) = 0.22 + 0.41 = 0.63$

Example
Consider the events A, B, and C within a sample space. You are given that the chance that A and B happens is 0.36, the chance that A and C happens is 0.22, and the chance that B and C occurs is 0.17. The probability that all 3 events happen is 0.06. In addition, $\Pr(\overline{A \cup B \cup C}) = 0$, $\Pr(A) = 0.6$, and $\Pr(B) = 0.55$. Find the chance that only C happens.

Solution:
The Venn diagram below aids in answering the question. The first regions to consider are the intersections between the sets. Knowing $\Pr(A \cap B \cap C) = 0.06$ helps to determine the chance of intersection between any two sets excluding the third set. For

example, $\Pr(A \cap B \cap \overline{C}) = \Pr(A \cap B) - \Pr(A \cap B \cap C) = 0.3$. Once you have the probability of each intersected region, you can use Pr(A) and Pr(B) to find $\Pr(A \cap \overline{B} \cap \overline{C})$ and $\Pr(\overline{A} \cap B \cap \overline{C})$, respectively. You are also given the fact that an outcome cannot lie outside the union of A, B, and C. That is, the probabilities in all the subregions within A, B, and C must sum to 1. As a result, the chance that just C happens is 0.21.

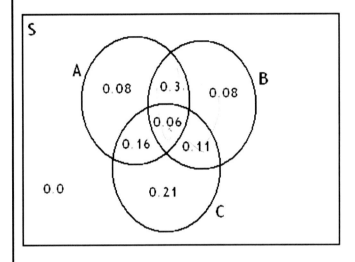

Section 1-4: Independent Events

Events may be dependent or independent with respect to each other. Events A and B are considered *independent* when the fact that A occurred or did not occur has no effect on the chances that B occurs or does not occur (and vice versa). Independent events are unconnected. Events that are independent have no influence on each other. On the other hand, events A and B are *dependent* when the observation that A happened (or did not happen) affects the probability that B happens.

Independence does not correspond to mutual exclusion. In fact, mutually exclusive events might be necessarily dependent upon each other. Consider events A and A^C. We know that only A or its complement must occur. If we are given that A happened, then we can know with 100% confidence that A^C did not happen.

A common illustration of dependence-independence appears when drawing marbles from a jar. When the process involves putting a drawn marble back in the jar every round before drawing again (called "with replacement"), previous draws will not affect the next draw, and the draws are independent. However, if the subject in the experiment does not replace each drawn marble (called "without replacement"), then the next draw will be dependent upon all the previous draws.

Suppose that Uriah has red and black marbles in a jar in certain proportions and draws marbles from the jar. If Uriah draws a marble and then puts it back in the jar, then separate drawings are independent. The next drawing will not be affected by previous drawings. However, suppose that Uriah does not replace the marbles. If he draws a red marble and does not replace it, then the next drawing will have a greater chance of producing a black marble. In this case, successive events are dependent.

The terms with/without replacement have broader implications in counting problems. If the problem allows elements to be reused in some organization, then whatever elements were used in the past do not influence future elements available. However, if elements cannot be reused, then the set of potential organizations is limited.

When two events A and B are independent, the chance that A and B both happen equals the product of the individual probabilities of the events.

Intersection of Independent Events Theorem
Events A and B are independent iff $P(A \cap B) = P(A) * P(B)$.

Example
Consider two farmers—Shem and Ezekiel. They live in opposite corners of the world and have no relation to each other. The probability that Shem plants wheat in his fields is 65%, and the chance that Ezekiel plants wheat is only 27%. What is the chance both farmers plant wheat?

Solution:
Let A = Shem plants wheat; P(A) = 0.65
Let B = Ezekiel plants wheat; P(B) = 0.27
It is given that A and B are independent.

P(A AND B) = $P(A \cap B) = P(A) * P(B) = 0.65 * 0.27 = 0.1755 = 17.55\%$

Example
The company TP Oil Industries will sign one of two contracts with window-washing companies A and B. Another company, Immobilized Oil Industries, is considering a contract with window-washing companies C and D. The probabilities of the options are shown below:

	Immobilized Oil Industries		
TP Oil Industries	Plan	C	D	Total
	A	0.1	0.15	0.25
	B	0.3	0.45	0.75
	Total	0.4	0.6	1.00

Are the decisions of the two oil companies independent?

Solution:
It is discovered that
$\Pr(A \cap C) = 0.1 = \Pr(A) * \Pr(C)$
$\Pr(A \cap D) = 0.15 = \Pr(A) * \Pr(D)$
$\Pr(B \cap C) = 0.3 = \Pr(B) * \Pr(C)$
$\Pr(B \cap D) = 0.45 = \Pr(B) * \Pr(D)$

Since all legal combinations of events follow the Intersection of Independent Events Theorem, the two oil companies are working independently from each other.

Example
A fish tank at a pet store has 6 fantail goldfish and 8 lionhead goldfish. A young child wants to buy 5 fish, and the store owner randomly nets 5 goldfish.
(a.) Find the chance that the child gets 5 fantail goldfish.
(b.) What is the probability that the child receives 1 fantail and 4 lionhead goldfish?

Solutions:
Netting fish from the tank is clearly a "without replacement" event because it's impossible for the child to get two copies of the same fish. All previous draws will impact available fish on the next draw.

(a.) Obtaining 5 fantails is equivalent to getting a fantail on the first draw, a fantail on the second draw knowing the first was a fantail, a fantail on the third draw knowing the first and second fish were fantails, and so on. Each draw of a fantail reduces the relative proportion of fantails in the tank.

$$\Pr(5 \text{ fantails}) = \frac{6}{14} * \frac{5}{13} * \frac{4}{12} * \frac{3}{11} * \frac{2}{10} = 0.002997$$

(b.)
The child must get the single fantail in one of five slots.
The probability that the first fish is a fantail and the other 4 are lionheads is:

$$\frac{6}{14} * \frac{8}{13} * \frac{7}{12} * \frac{6}{11} * \frac{5}{10} = \frac{6}{143}$$

The probability that the second fish is the fantail and the others are lionheads is:

$$\frac{8}{14} * \frac{6}{13} * \frac{7}{12} * \frac{6}{11} * \frac{5}{10} = \frac{6}{143}$$

Note that the placement of the fantail does not affect the probability of getting just one fantail. The numerators and denominators contain exactly the same terms (in different orders) multiplied together.

The final probability of 1 fantail and 4 lionheads is:

$$\Pr = 5 * \frac{6}{143} = \frac{30}{143}$$

Section 1-5: Conditional Probability

The universe is incredibly interconnected. No event or object stands alone, unaffected by the activities of other bodies. The probability of an outcome can be greatly influenced by whether or not another event happened. For instance, given that a child's parents smoke, the child has an increased chance of starting the habit at a later age. Under conditional probability, the chance of some event is related to another event in the past. If we know that event A happened, the probabilities for other events may be changed. In other words, an event could be a predictor of other events. Conditional probability assumes dependence between events. If events truly are independent, then the events will not affect each other's probability.

The conditional operator is a vertical segment, like | . The symbol is read "given." For instance, the expression Pr(A | B) reads "the probability of A given B." The event on the right side of the operator already happened. To compute a conditional probability, follow the formula below. The given event is in the denominator.

Conditional Probability Formula
$$\Pr(A \mid B) = \frac{\Pr(A \cap B)}{\Pr(B)}$$

The basic conditional probability formula can have its terms slightly rearranged in the succeeding corollaries:

Conditional Probability Corollaries
(1.) $P(A \cap B) = P(A \mid B) * P(B)$
(2.) $P(A \cap B) = P(B \mid A) * P(A)$

A perplexing case of the conditional probability formula happens when you are given that event B happened even though $\Pr(B) = 0$. In this case, $\Pr(A \mid B)$ would be undefined.

Example
Fred is conducting a study on obesity. He notices a relationship between access to a big-box wholesale store and rapid weight gain. Given that a subject has access to a wholesale store, the probability that the person is obese is 0.43. 70% of people in the study had access to a wholesale store. What is the chance that a randomly chosen person from the study is obese and lives near a wholesale store?

Solution:
Let A = obese, B = access to wholesale

P(A | B) = 0.43
P(B) = 0.70
We need to find $P(A \cap B)$

$$P(A|B) = \frac{P(A \cap B)}{P(B)} \rightarrow 0.43 = \frac{P(A \cap B)}{0.70} \rightarrow P(A \cap B) = 0.301$$

Example
Suppose P(A) = 0.28, P(B) = 0.51, and P(A | B) = 0.17. Find P(B | A).

Solution:
$$P(B|A) = \frac{P(A \cap B)}{P(A)} = \frac{P(A \cap B)}{0.28}$$

We need to find $P(A \cap B)$. Since we already have P(A | B), we can use the formula $P(A \cap B) = P(A|B) * P(B)$.

$P(A \cap B) = 0.17 * 0.51 = 0.0867$

Now, $P(B|A) = \dfrac{0.0867}{0.28} = 0.3096$

The complement operator can apply to events under conditional probability. For example, if you computed Pr(A | B), you can uncover $\Pr(\overline{A} | B)$ by calculating $1 - \Pr(A|B)$.

If the events of interest are independent, then the fact that one event occurred will not affect the chances of other events. You can remove the "given" operator and the "given" operand from probability expressions when the events are independent.

Conditional Probabilities under Independence
Suppose that events A and B are independent. Then,
(1.) Pr(A | B) = Pr(A)
(2.) Pr(B | A) = Pr(B)

Example
Hans, an aerospace engineer, is examining data from recent airplane crashes. He built the table below, where each cell gives a count of crashes:

	superjet type A	superjet type B	Total
pilot error	36	126	**162**
mechanical problem	72	252	**324**
bad weather	18	63	**81**
Total	**126**	**441**	**567**

(a.) Hans picks one crash at random to examine in fine detail. What is the chance that the plane is a superjet type B? What is the probability the plane crashed in bad weather?
(b.) Is the jet type independent from the cause of crash?

Solutions:
(a.) Pr(superjet type B) = 441 / 567 = 7/9
 Pr(bad weather) = 81 / 567 = 1/7

(b.) A quick way to tell if the two types of events are independent from each other is to determine if all the columns are proportional to each other and all the rows are proportional to each other. The column for "superjet type B" equals the column for "superjet type A" multiplied by 3.5. The rows are likewise proportional to each other. Therefore, jet type is independent from crash cause.

A more detailed analysis of whether jet type and crash cause are independent involves computing the probabilities of conditional events. We shall examine just a subset of the possible conditional probabilities. Let

C_1 = pilot error, C_2 = mechanical problem, C_3 = bad weatner

$Pr(C_1) = 2/7$, $Pr(C_2) = 4/7$, $Pr(C_3) = 1/7$

$Pr(C_1 \mid$ superjet type A$) = Pr(C_1$ and superjet type A$) / Pr($superjet type A$)$
 $= (36/567) / (126/567) = 2 / 7 = Pr(C_1)$
$Pr(C_2 \mid$ superjet type B$) = Pr(C_2$ and superjet type B$) / Pr($superjet type B$)$
 $= (252/567) / (441/567) = 4/7 = Pr(C_2)$
………..

$Pr($superjet type A $\mid C_3) = Pr($superjet type A and $C_3) / Pr(C_3)$
 $= (18 / 567) / (81 / 567) = 2 / 9 = Pr($superjet type A$)$
………..

After finding all possible conditional probabilities, it should appear that $Pr(A \mid B) = Pr(A)$ and $Pr(B \mid A) = Pr(B)$, where A is the jet type and B is the crash cause. Therefore, jet type and crash cause are independent.

Example
Abram conducts a survey on political views and collects the tabulated data below:

	socialist	capitalist	Total
poor	115	40	155
rich	10	50	60
Total	125	90	215

Each interviewee categorized himself/herself as rich or poor and as a socialist or capitalist.

Find the probability that a chosen subject is....
(a.) poor   (b.) a capitalist   (c.) rich and a capitalist
(d.) poor given that the individual is a socialist
(e.) a capitalist given that the person is rich

Solutions:
(a.) The experiment used 215 people. Assume that each person in the study has the same chance of being chosen.

$$P(\text{poor}) = \frac{155}{215} = 0.7209$$

(b.) $P(\text{capitalist}) = \dfrac{90}{215} = 0.4186$

(c.) $P(\text{rich AND capitalist}) = \dfrac{50}{215} = 0.2326$

(d.) $P(\text{poor} \mid \text{socialist}) = \dfrac{P(\text{poor} \cap \text{socialist})}{P(\text{socialist})} = \dfrac{115/215}{125/215} = \dfrac{115}{125} = 0.92$

(e.) $P(\text{capitalist} \mid \text{rich}) = \dfrac{P(\text{capitalist} \cap \text{rich})}{P(\text{rich})} = \dfrac{50/215}{60/215} = \dfrac{50}{60} = 0.8333$

If a problem has many possible events in the same related class, you might consider building a conditional probability tree. The typical conditional probability tree has a root node on the far left side. It represents the "start" of the analysis. The researcher can then choose one of a few potential routes. Each possible route creates a branch stemming from the root. At each node of the tree, the researcher (or just fate) will choose another valid action. The branching factor at each node is the number of possible options or events at that node. The depth of the tree equals the maximum number of consecutive decisions that could be made on a particular route.

Example

A video game company is considering creating a new title. Let event M be that the game sells at least one million copies. The company can choose from several genres of games. The probability that it will choose genre A_i is given below:

Event A_i	$P(A_i)$
A_1 = action game	0.65
A_2 = role playing game	0.05
A_3 = strategy game	0.30

The company will make the choice of game randomly. This is the first act and will create three potential branches from the root.

Given that the company chose game A_i, the probability of M is:

Event	P(Event)
$M \mid A_1$	0.2
$M \mid A_2$	0.1
$M \mid A_3$	0.6

(a.) Find $P(A_i \cap M)$ and $P(A_i \cap \overline{M})$ for $1 \leq i \leq 3$.

(b.) Find $P(M)$ and $P(\overline{M})$.

Solutions:

(a.) First, we should build a probability table for $P(\overline{M} \mid A_i)$. These probabilities are complements of $P(M \mid A_i)$.

Event	P(Event)
$\overline{M} \mid A_1$	0.8
$\overline{M} \mid A_2$	0.9
$\overline{M} \mid A_3$	0.4

We need to find probabilities for the events in the rightmost column. The diagram is a bit busy with events labeled at the nodes and probabilities along the joining segments. Note that the branching factor after each A_i is 2 since the children are M and $\overline{M}$.

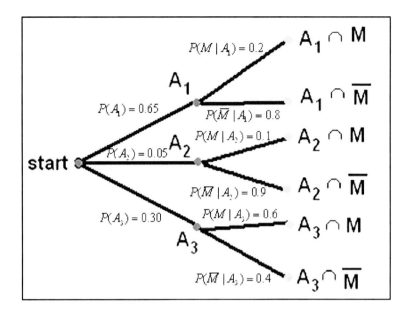

To compute $P(A_1 \cap M)$, simply multiply $P(A_1)$ with $P(M \mid A_1)$. A similar rule holds for complement-M. We are using the conditional probability corollaries.

$P(A_1 \cap M) = P(A_1) * P(M \mid A_1) = 0.65 * 0.2 = 0.13$
$P(A_1 \cap \overline{M}) = P(A_1) * P(\overline{M} \mid A_1) = 0.65 * 0.8 = 0.52$

$P(A_2 \cap M) = P(A_2) * P(M \mid A_2) = 0.05 * 0.1 = 0.005$
$P(A_2 \cap \overline{M}) = P(A_2) * P(\overline{M} \mid A_2) = 0.05 * 0.9 = 0.045$

$P(A_3 \cap M) = P(A_3) * P(M \mid A_3) = 0.3 * 0.6 = 0.18$
$P(A_3 \cap \overline{M}) = P(A_3) * P(\overline{M} \mid A_3) = 0.3 * 0.4 = 0.12$

Notice that the probability of A_1 is 0.65. We divide the 65% between $(A_1 \cap M)$ and $(A_1 \cap \overline{M})$. A similar pattern holds for A_2 and A_3.

(b.) $P(M) = P(A_1 \cap M) + P(A_2 \cap M) + P(A_3 \cap M) = 0.13 + 0.005 + 0.18 = 0.315$
$P(\overline{M}) = 1 - P(M) = 1 - 0.315 = 0.685$
$\phantom{P(\overline{M})} = P(A_1 \cap \overline{M}) + P(A_2 \cap \overline{M}) + P(A_3 \cap \overline{M}) = 0.52 + 0.045 + 0.12 = 0.685$

Adding conditional event(s) to a probability statement is incredibly easy. All the given events which have occurred are placed to the right of the vertical bar in every probability expression. As an example, suppose we are concerned about whether event A happens, and we already know events B and C happened. The basic conditional probability statement could be written as:

$$\Pr(A \mid B \cap C) = \frac{\Pr(A \cap B \cap C)}{\Pr(B \cap C)}$$

Prior and Posterior Probabilities

A distinction exists between a prior and posterior probability. A *prior probability* is the probability of an event before any extra information is given. An example of a prior probability is $\Pr(A)$. Notice that the "given" symbol is not used in a prior probability. On the other hand, a *posterior probability* describes the chance of an event after you know that some other event happened. The posterior probability of A knowing that B happened is written as $\Pr(A \mid B)$.

Example
Tyrone is gambling in the street. He first spins an equally balanced, three-sided top with the colors red, green, and blue. Depending on the color of the top, he selects a chip from one of three hats also colored red, green, and blue. The chip selected shows his payoff/loss. The red hat has 10 chips labeled "-2" and 5 labeled "+1." The green hat has 3 chips labeled "-2" and 7 labeled "+1." The blue hat has 8 chips labeled "-2" and 4 labeled "+1."

Find the chance that Tyrone gets a positive payoff in the next round.

Solution:
The probability of a positive payoff from a hat depends on its color.

Pr(positive payoff) = Pr(red top AND +1 from red hat) +
  + Pr(green top AND +1 from green hat) + Pr(blue top AND +1 from blue hat)

Pr(red top AND +1 from red hat) = Pr(red top) * Pr(+1 from red hat | red top)
$$= \frac{1}{3} * \frac{5}{15} = \frac{1}{9}$$

Pr(green top AND +1 from green hat) = Pr(green top) * Pr(+1 from green hat | green top)
$$= \frac{1}{3} * \frac{7}{10} = \frac{7}{30}$$

Pr(blue top AND +1 from blue hat) = Pr(blue top) * Pr(+1 from blue hat | blue top)
$$= \frac{1}{3} * \frac{4}{12} = \frac{1}{9}$$

$$\Pr(\text{positive payoff}) = \frac{1}{9} + \frac{7}{30} + \frac{1}{9} = \frac{41}{90}$$

Section 1-6: Law of Total Probability and Bayes' Theorem

The Law of Total Probability provides a method to partition an event into smaller parts and find the probability of the main event by summing the probabilities of each part.

Law of Total Probability
Let the sets B_1, B_2, ..., B_n compose a partition of S. Let A be an event. The sets $(B_1 \cap A)$, $(B_2 \cap A)$, ..., $(B_n \cap A)$, formed by intersecting A with each of the B_i's, compose a partition of A. Then,

$$\Pr(A) = \sum_{i=1}^{n} \Pr(B_i \cap A) = \sum_{i=1}^{n} \Pr(A \mid B_i) \bullet \Pr(B_i)$$

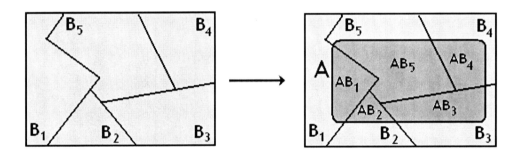

As seen in the diagram above, AB_i and AB_j are disjoint for all $i \neq j$. The Law of Total Probability also says that
$$A = (B_1 \cap A) \cup (B_2 \cap A) \cup ... \cup (B_n \cap A)$$

Example
Government investigators are pursuing a computer hacker. They believe the hacker could be in San Francisco, Chicago, or Miami. The chance the hacker is in each city is:

Pr(SF) = 52%, Pr(C) = 18%, Pr(M) = 30%

The government is concerned the hacker will try to break into a bank and take its assets. Let A denote the event that the hacker steals the assets. The government believes the ability for the hacker to steal the assets depends on the hacker's location. It has formed the following probabilities of theft given the hacker's location:

$\Pr(A \mid SF) = 80\%, \quad \Pr(A \mid C) = 45\%, \quad \Pr(A \mid M) = 24\%$

What is the chance the hacker steals the assets?

Solution:
The hacker can be in only one city, so the events SF, C, and M are disjoint. Therefore, the events $A \cap SF$, $A \cap C$, and $A \cap M$ are also disjoint (the hacker could not be in multiple cities at the same time stealing money).

Using the law of total probability,
$$\Pr(A) = \Pr(A \mid SF) * \Pr(SF) + \Pr(A \mid C) * \Pr(C) + \Pr(A \mid M) * \Pr(M)$$
$$= 0.80 * 0.52 + 0.45 * 0.18 + 0.24 * 0.30$$
$$= 0.569$$

Example
A life insurance company is examining an insurance claim for validity. Some insureds fake their deaths to collect the insurance benefit. The chance of a claim being fraudulent depends on the state-declared cause of death. Potential causes of death include drowning (D), building fire (F), and vanishing while hiking (H). Let 'I' signify an invalid claim. The financial analysts developed these probabilities:

$\Pr(I \mid D) = 12\%, \quad \Pr(I \mid F) = 37\%, \quad \Pr(I \mid H) = 85\%$

The probability of any insured "dying" from the possible causes are:

$\Pr(D) = 16\%, \quad \Pr(F) = 7\%, \quad \Pr(H) = 77\%$

The insurance company receives a claim from a beneficiary but does not yet know the cause of "death." What's the chance this claim is invalid?

Solution:
Let's assume the causes of death are disjoint so that no one could have the misfortune of dying in multiple terrible ways.

$$\Pr(I) = \Pr(I \mid D) * \Pr(D) + \Pr(I \mid F) * \Pr(F) + \Pr(I \mid H) * \Pr(H)$$
$$= 0.12 * 0.16 + 0.37 * 0.07 + 0.85 * 0.77$$
$$= 0.6996$$

Bayes' Theorem is a convenient formula for finding a conditional probability, $\Pr(B_i \mid A)$, assuming you already know all the opposite conditional probabilities, $\Pr(A \mid B_j)$. Bayes' Theorem is essentially an expanded form of the conditional probability equation.

Bayes' Theorem

Suppose that set B contains n possible events labeled $B_1, B_2, \ldots, B_n$. All the B_i events must be disjoint, and they partition S. Let A be another event with positive probability. Then,

$$\Pr(B_i \mid A) = \frac{\Pr(A \mid B_i) * \Pr(B_i)}{\sum_{j=1}^{n} \Pr(A \mid B_j) * \Pr(B_j)} = \frac{\Pr(A \cap B_i)}{\Pr(A)}$$

Note that the denominator in Bayes' Theorem is just $\Pr(A)$ taken from the Law of Total Probability.

Example

A deck of cards has 4 red cards labeled 1-4, 7 white cards labeled 1-7, and 5 blue cards labeled 1-5. The experimenter draws one card, and the value is greater than 3. What is the probability for each card color?

Solution:
Let B_1 = red card drawn, $\Pr(B_1) = 4/16$
Let B_2 = white card drawn, $\Pr(B_2) = 7/16$
Let B_3 = blue card drawn, $\Pr(B_3) = 5/16$
Let A = the card value is greater than 3.

$\Pr(A \mid B_1) = 1/4$, $\Pr(A \mid B_2) = 4/7$, $\Pr(A \mid B_3) = 2/5$

$$\begin{aligned}\Pr(A) &= \Pr(AB_1) + \Pr(AB_2) + \Pr(AB_3) \\ &= \Pr(A \mid B_1)*\Pr(B_1) + \Pr(A \mid B_2)*\Pr(B_2) + \Pr(A \mid B_3)*\Pr(B_3) \\ &= (1/4)*(4/16) + (4/7)*(7/16) + (2/5)*(5/16) \\ &= (4/64) + (28/112) + (10/80) \\ &= 7/16 \end{aligned}$$

$$\Pr(B_1 \mid A) = \frac{\Pr(AB_1)}{\Pr(A)} = \frac{4/64}{7/16} = \frac{1}{7}$$

$$\Pr(B_2 \mid A) = \frac{\Pr(AB_2)}{\Pr(A)} = \frac{28/112}{7/16} = \frac{4}{7}$$

$$\Pr(B_3 \mid A) = \frac{\Pr(AB_3)}{\Pr(A)} = \frac{10/80}{7/16} = \frac{2}{7}$$

Example
Jed and Gunner are two auto mechanics. When a customer brings in a vehicle for repair, the boss assigns the car to either Jed or Gunner. The probability that Jed works on the car is 55%. Assuming that Gunner worked on the car, the probability that it now runs is 80%. Given that Jed worked on the vehicle, the chance it doesn't work is 35%.

(a.) Find the probability that the car runs after repairs.
(b.) What is the chance that Gunner worked on the car, given that it runs now?
(c.) What is the chance that Jed worked on the car, given that it doesn't run now?

Solutions:
(a.)
Let J = Jed works on the car, G = Gunner works on the car, R = the car runs after the fix

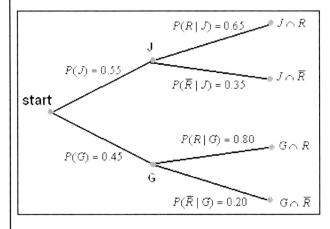

We should first find the probabilities in the rightmost column.
$P(J \cap R) = P(J) * P(R \mid J) = 0.55 * 0.65 = 0.3575$
$P(J \cap \overline{R}) = P(J) * P(\overline{R} \mid J) = 0.55 * 0.35 = 0.1925$
$P(G \cap R) = P(G) * P(R \mid G) = 0.45 * 0.80 = 0.36$
$P(G \cap \overline{R}) = P(G) * P(\overline{R} \mid G) = 0.45 * 0.20 = 0.09$

$P(R) = P(J \cap R) + P(G \cap R) = 0.3575 + 0.36 = 0.7175$

(b.)
There are two ways to solve this problem. You can use the basic conditional probability equation:

$$P(G \mid R) = \frac{P(G \cap R)}{P(R)} = \frac{0.36}{0.7175} = 0.5017$$

…or you can use Bayes' Theorem:

$$P(G \mid R) = \frac{P(R \mid G) * P(G)}{P(R \mid G) * P(G) + P(R \mid J) * P(J)} = \frac{0.80 * 0.45}{0.80 * 0.45 + 0.65 * 0.55} = \frac{0.36}{0.36 + 0.3575}$$
$$= 0.5017$$

(c.)
We first need the probability that the car doesn't run after repairs:
$$P(\overline{R}) = 1 - P(R) = 1 - 0.7175 = 0.2825$$

$$P(J \mid \overline{R}) = \frac{P(J \cap \overline{R})}{P(\overline{R})} = \frac{0.1925}{0.2825} = 0.6814$$

Check using Bayes' Theorem:
$$P(J \mid \overline{R}) = \frac{P(\overline{R} \mid J) * P(J)}{P(\overline{R} \mid J) * P(J) + P(\overline{R} \mid G) * P(G)} = \frac{0.35 * 0.55}{0.35 * 0.55 + 0.20 * 0.45}$$
$$= \frac{0.1925}{0.1925 + 0.09} = 0.6814$$

In this usage of Bayes' Theorem, the variable of interest is the person who fixed the car, represented by either event J or G. We must allow this variable to vary in the right side of the equation.

Chapter 2: Counting Tools

Many probability problems can be easily solved if you know how many outcomes exist within events of interest. Typically, each outcome is composed of a tuple of elements. The tuple might need to be organized in a certain fashion so that order matters, or perhaps order doesn't matter for the sub-elements. For example, if the experiment involves selecting 3 students from a class of 20 for an extra credit assignment, we clearly can't repeat students in the selection, and the order in which we pick them might not matter. Each tuple would be the list of students selected, such as <Amir, Uriah, DeShawn>. Many experiments have an extremely large number of ways in which they can be executed. Rather than explicitly counting every element in a collection, you can use shortcut methods to determine how many elements exist. The next few sections cover factorials, permutations, combinations, and even multinomial combinations. All these tools help to solve problems with the first probability distribution—the discrete uniform distribution. This chapter gives the basics of combinatorial analysis—an entire branch of math dealing with counting objects.

Section 2-1: Counting by Multiplication

Factorials

Recall that the factorial operator means to multiply all integers from the starting argument down to 1, inclusive. That is, $x! = x*(x-1)*(x-2)*...*2*1$.

Factorial Properties
(1.) $n! = n*(n-1)!$
(2.) $0! = 1$
(3.) $(n-a)! = (n-a)*(n-a-1)!$
(4.) $\dfrac{n!}{(n-a)!} = n*(n-1)*(n-2)*....*(n-a+1)$, where 'a' is a positive integer

---

Forming a Tuple of Objects from Multiple Categories

In many situations, the experimenter will need to select one object from each available category. For instance, a college student might need to pick one history class from a set of 15 potential classes, an English class from 4 options, and a math class from 25 options. The selections form a tuple, such as <British History, Poetry, and Calculus>. The amount of options in one category must be independent from the amount of options in all other categories. Although the choice made in one category might affect the actual range of choices in future categories, the number of options in each category must be unaffected by a previous choice. To find the total quantity of tuples which can be constructed by taking one object from each category, simply multiply together the sizes of all categories.

---

mn Rule
Suppose you have m options in one set labeled $a_1, a_2, ..., a_m$ and n options in a second set labeled $b_1, b_2, ..., b_n$.

The total number of pairs using one object from the first set and another object from the second set is m*n.

Number of Possible Selections of Objects in m Dimensions
Consider an experiment which will produce an outcome containing m dimensions. Each dimension could be a value chosen from a set of available values or an object taken from a collection of objects. Let the first dimension have n_1 possible values, the second dimension have n_2 possible values, and so on. The number of ways to select the m-tuple outcome is:

$$n_1 * n_2 * ... * n_m$$

Section 2-2: Permutations and Combinations

A *permutation* is a particular ordering of objects. Order matters in a permutation. To permute two objects means to switch them by location in a sequence. Permutations are easily seen in the letters forming words. If you mix up the letters for a word, you of course obtain gibberish or a completely different word with a new meaning. Permutations are without replacement processes; once you use an element, you cannot reuse it.

Basic Permutation Theorem
Given n distinct elements, there exist n! permutations using all the elements.

Example
How many ways can you arrange the letters A through E, using each letter exactly once?

Solution:
Order matters in this problem, so permutations are needed. You can solve the problem using the basic permutation theorem, or you can think through the underlying principles.

We need to form a 5-letter "word" using each letter precisely once. Imagine that we have five slots with indices below them:

```
┌──┬──┬──┬──┬──┐
│  │  │  │  │  │
└──┴──┴──┴──┴──┘
 0  1  2  3  4
```

Slot #0 can take one of five available letters. Then, slot #1 has four possible letters left. Slot #2 has three possible values, slot #3 has two values, and slot #4 will be forced to take the last remaining letter. Multiply the quantities of available values for each slot:

Total number of permutations = $5*4*3*2*1 = 5! = 120$

Permutations have a simple mathematical notation. To designate a permutation, write $_nP_r$. The 'P' stands for "permutation," the letter 'n' shows how many objects are in the full pool, and the letter 'r' reveals the quantity of objects in each subset. The expression $_nP_r$ reads "the number of permutations of n objects taken r at a time." It gives the amount of ways to order r elements from a bigger pool of n elements.

Permutation Formula
$$_nP_r = \frac{n!}{(n-r)!} = n*(n-1)*(n-2)*...*(n-r+1)$$

Example
The Lake Conference has 9 math teams competing. How many organizations exist for the top two teams?

Solution:
n = 9, r = 2 (we are concerned about rank #1 and rank #2 only)

Since order clearly matters, we should use permutations.
$$_nP_r = \frac{n!}{(n-r)!} = \frac{9!}{(9-2)!} = \frac{9!}{7!} = \frac{9*8*7*6....}{7*6*....} = 9*8 = 72$$

Example
The state prison system is designing new prisoner ID's. The administration wants to know how many ID's are possible using the format:
(three unique letters taken from A-Z), (six repeatable digits taken from the values 0-4)

Solution:
The order of symbols matters for ID tags, so permutations are required.
For the letters, n = 26 and r = 3.
$$\text{Then, } _{26}P_3 = \frac{26!}{(26-3)!} = \frac{26!}{23!} = \frac{26*25*24*23*22*...}{23*22*...}$$
$$= 26*25*24 = 15,600$$

The digits can be repeated, so each slot in the numerical portion could take one of five values.
    quantity of numbers = 5 * 6 = 30

Total number of ID's = 15,600 * 30 = 468,000 (enough to accommodate prisoners for the next month)

Example
Seven employees are vying for the positions of senior executive v.p., executive v.p., and junior v.p. Assuming that all the employees are equally obedient yes-men and play golf with the top boss in equal proportions, how many arrangements of the employees are possible?

$$_7P_3 = \frac{7!}{(7-3)!} = \frac{7*6*5*4*3*2*1}{4*3*2*1} = 7*6*5 = 210$$

A *combination* is a selection of elements from a larger pool. Order does not matter in a combination. As an example, suppose that 50 basketball players are divided into separate teams of 10 players each. Once a team is formed, it does not matter which player was added to the team first as all the players will mix together. Combinations are also without replacement processes; an element cannot be reused.

The mathematical notation for a combination is $_nC_r$ or $\binom{n}{r}$. The expression $_nC_r$ is pronounced "n choose r" or "the number of combinations of n objects taken r at a time."

Combination Formula

$$_nC_r = \binom{n}{r} = \frac{n!}{(n-r)! \cdot r!}$$

Example

Suppose Ezra is an accounting student who must take 4 elective classes out of 13 possible classes. The classes have no prerequisites, so he can take them in any order. In how many ways can he select courses?

Solution:
n = 13, r = 4

$$_nC_r = {_{13}C_4} = \frac{n!}{(n-r)! \cdot r!} = \frac{13!}{(13-4)! \cdot 4!} = \frac{13!}{9! \cdot 4!} = \frac{13 \cdot 12 \cdot 11 \cdot 10}{24} = 715$$

Example

The state hockey association has 40 teams. Assuming that all teams start equally strong, and disregarding which teams play each other, how many combinations of teams exist among the final top four?

$$_{40}C_4 = \frac{40!}{(40-4)! \cdot 4!} = \frac{40!}{36! \cdot 4!} = \frac{40 \cdot 39 \cdot 38 \cdot 37}{24} = 91{,}390$$

Example

A village is holding a military draft for an important mission. It has 15 young men to choose from, but only needs 5 soldiers. How many ways exist to assemble the team?

Solution:

$$_{15}C_5 = \frac{15!}{(15-5)! \cdot 5!} = \frac{15!}{10! \cdot 5!} = \frac{15 \cdot 14 \cdot 13 \cdot 12 \cdot 11}{120} = 3{,}003$$

Example

An ice cream shop has 25 unique toppings. A customer can use 4 toppings. Suppose a customer randomly picks four toppings without repeating a topping. What is the chance he does not choose any of the following: nuts, pure sugar, and sprinkles?

Digital Actuarial Resources   *Comprehensive Probability Review for Actuarial Exams*

Solution:
First, find how many combinations are possible for all 25 toppings.
$$_{25}C_4 = 12,650$$

Second, find the number of combinations using just the 22 "allowed" toppings.
$$_{22}C_4 = 7,315$$

The final probability equals the fraction:
$$\frac{(\# \text{ combinations of toppings that exlude the listed 3})}{(\text{total } \# \text{ of combinations})}$$
$$= \frac{7,315}{12,650} = 0.57826$$

Combination Properties
(1.) $_nC_{n-1} = n$
(2.) $_{n+1}C_r = {_nC_{r-1}} + {_nC_r}$
(3.) $_nC_r = {_nC_{(n-r)}}$

The permutation and combination formulas can be connected with: $_nP_r = {_nC_r} * r!$
The combination function is symmetric in the sense that it produces equal values depending on the relative magnitude of r.

To summarize, permutations and combinations are both "without replacement" calculations. The only difference between a permutation and combination is whether or not the order of selected elements matters. Since order matters in a permutation but not in a combination, the quantity of permutations is generally larger than the amount of combinations using elements from a pool.

Example
A military captain must divide 15 soldiers into two teams. One team will have 9 members and will storm the hill, while the second team has 6 members and will take the beach. The captain will pick teams randomly. Hans and Klaus are two friends in the military. Find the chance that Hans and Klaus are on the same team.

Solution:
First, find the total number of ways to partition the 15 soldiers into the two teams. Note that order doesn't matter; a soldier in the group storming the hill will not care about whether he was picked first or last. Consequently, let's use combinations for counting.

Total number of partitions = $_{15}C_9 = \dfrac{15!}{6!*9!} = 5,005$

Number of ways in which Hans and Klaus are both in the team storming the hill:
$$_{13}C_7 = \frac{13!}{6!*7!} = 1,716$$
(2 people—Hans and Klaus—are already guaranteed to be in the group storming the hill. 13 soldiers remain, of which 7 must be placed in the first group.)

Number of ways in which Hans and Klaus are both in the team taking the beach:
$$_{13}C_4 = 715$$

Since the two friends could not both be in the team storming the hill and in the team taking the beach, we can add the number of outcomes in these mutually exclusive events.

The final probability is:
$$\Pr = \frac{1,716 + 715}{5,005} = 0.48571$$

Example
A neighborhood association has 30 home owners. Every year, 5 owners are selected randomly to run the neighborhood. The assigned positions are important because they include: the president, the vice president, the code enforcer, the fee collector, and the lawn maintenance person. Three neighbors—Axel, Dredge, and Uriah—have a plan to overturn the association, but two of them must be president and vice president, and the third person must have one of the remaining jobs. What is the chance that the rebels can break up the association within the next year?

Solution:
Since positions in the association are unique, the order of selected people is important. Let's use permutations in counting.

Total number of associations = $_{30}P_5 = \frac{30!}{25!} = 17,100,720$

The potential orderings of Axel, Dredge, and Uriah in strategic positions include:
(A, D, U, -, -); (A, D, -, U, -); (A, D, -, -, U)
(A, U, D, -, -); (A, U, -, D, -); (A, U, -, -, D)
(D, A, U, -, -); (D, A, -, U, -); (D, A, -, -, U)
(D, U, A, -, -); (D, U, -, A, -); (D, U, -, -, A)
(U, A, D, -, -); (U, A, -, D, -); (U, A, -, -, D)
(U, D, A, -, -); (U, D, -, A, -); (U, D, -, -, A)

Each dash can be any of the other 27 people.

Each ordered pair has $1*1*1*27*26 = 702$ possible ways of happening. With 18 ordered pairs, there are $702*18 = 12,636$ ways that the three neighbors can overthrow the association.

$$\Pr(\text{overthrow}) = \frac{12{,}636}{17{,}100{,}720} = 0.0007389$$

Section 2-3: Multinomial Combinations

Multinomial coefficients provide a way to compute the number of combinations of n elements that must be placed into k groups, where $k \geq 2$. Although the objects are traditionally quite similar, each of the n elements is viewed as a distinct entity. The ordering for objects within a group does not matter. The number of objects that must fit in the ith class is denoted as x_i. Each class count must be a positive integer. Additionally, all the classes are mutually exclusive; an object must belong in precisely one class. All n elements must be placed into groups so that $x_1 + x_2 + \ldots + x_k = n$.

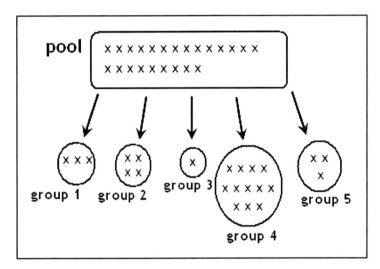

The multinomial coefficient formula depends on the factorial operator. The formula gives the number of possible ways to classify n objects in k available groups. Within parentheses, the total number of objects to place into classes is denoted 'n' and exists at the top. The individual class sizes exist at the bottom of the parentheses and are separated by commas. To compute a multinomial coefficient, transfer each term to a fraction, find the factorial of each term, and multiply the factorials together.

Multinomial Coefficient Formula

$$\binom{n}{x_1, x_2, \ldots x_k} = \frac{n!}{x_1! \cdot x_2! \cdot \ldots \cdot x_k!}$$

The expression above reads "n choose $x_1, x_2, \ldots, x_k$."

Example

A manager at a retail store has 14 new employees which he must place in departments. He needs 5 employees in the garden center, 3 in the auto lube, 4 in plumbing, and 2 in the junk food aisle. Assuming that all employees are at the same status and the order of their placement doesn't matter, how many ways can the manager form the departments?

Solution:
$n = 14, x_1 = 5, x_2 = 3, x_3 = 4, x_4 = 2$

$$\binom{14}{5, 3, 4, 2} = \frac{14!}{5! \cdot 3! \cdot 4! \cdot 2!} = 2{,}522{,}520$$

Example

A disgruntled airline worker is randomly throwing bags onto planes. He has 20 pieces of luggage and randomly chooses 3 bags for the flight to Phoenix, 7 bags for Detroit, 1 bag for Minneapolis, and 9 bags to Newark. In how many ways can the luggage handler assign bags to airplanes?

Solution:

$$\binom{20}{3, 7, 1, 9} = \frac{20!}{3! \cdot 7! \cdot 1! \cdot 9!} = 221{,}707{,}200$$

Example

A syndicate of 200 subversives is attempting to influence the regional government. The syndicate forms 3 smaller groups of lobbyists. Group A will contain 3 people and will bribe the government to reduce regulation of banks. Group B will hold 4 people and will encourage the government to allow more oil drilling. Group C will have 5 people and will push for the legalization of all firearms. Assume that members of the syndicate are randomly assigned to the subgroups, and a person could belong to at most one subgroup. Calculate the number of potential ways in which the subgroups could be formed.

Solution:
There are actually four groups of the following sizes:
$x_1 = 3, x_2 = 4, x_3 = 5$, and $x_4 = 188$.
The fourth "group" consists of all people not belonging to a special lobbyist group.

Total number of subgroup formations =
$$\binom{200}{3, 4, 5, 188} = \frac{200!}{3! * 4! * 5! * 188!} = \frac{200 * 199 * \ldots * 189}{3! * 4! * 5!} = 1.69305 * 10^{23}$$

Example
A variety of territorial rodents have occupied a 16-unit apartment complex. Each unit is overtaken by one species of rodent. Due to natural forces, the rodents have filled the complex in a random fashion. 6 rooms have raccoons, 8 rooms have rats, and 2 rooms have squirrels. Humans cautiously live in 9 of the available units. What is the chance that among the rooms occupied by humans, 3 have raccoons, 5 have rats, and 1 room has squirrels?

Solution:
First, let's find the total number of ways in which the rodents could spread themselves among the rooms.

$$T = \binom{16}{6, 8, 2} = \frac{16!}{6! * 8! * 2!} = 360,360$$

9 rooms have humans, and the other 7 rooms don't have people. An animal must either be in a human-filled room or in a room without people. The number of ways of dividing the human-occupied rooms in such as fashion that 3 have raccoons, 5 have rats, and 1 has squirrels is:

$$A = \binom{9}{3, 5, 1} * \binom{7}{3, 3, 1} = \frac{9!}{3! * 5! * 1!} * \frac{7!}{3! * 3! * 1!} = 504 * 140 = 70,560$$

$$\text{final probability} = \frac{70,560}{360,360} = 0.1958$$

The binomial coefficient is a special case of the general multinomial coefficient with k = 2. When one category has x elements, the second category must have (n – x) elements. The multinomial coefficient is

$$\binom{n}{x, n-x} = \frac{n!}{x! \cdot (n-x)!}$$

which is really $_nC_x$.

Section 2-4: Discrete Uniform Distribution

The simplest and most fundamental probability distribution for events is the discrete uniform distribution. This distribution is also called the *counting distribution* or the *simple sample space distribution*. It applies to a sample space and events with finitely many outcomes; the quantity of outcomes cannot be infinity. Each outcome has the same chance of occurring. If a sample space has n elements, then the chance of each outcome is 1/n. During a run of the experiment, the outcome is chosen randomly, not giving bias to any element. To compute the probability of an event, simply count the number of outcomes that lie in the event and divide by the total number of outcomes in the sample space.

Simple Sample Space Probability Formula
$$\Pr(A) = \frac{N(A)}{N(S)} = \frac{(\#\text{ outcomes in A})}{(\#\text{ outcomes in S})}$$

You need to be consistent in how you count the elements in the numerator and denominator. You need to decide whether or not order of the elements matters for each outcome and then use the same system for counting elements in A and in S. That is, if you use combinations in the denominator, you should also use combinations in the numerator. If you use permutations in the numerator, permutations will probably also exist in the denominator.

Example
Erwin has a deck of 24 playing cards. 12 cards are red and 12 are black. He deals 8 cards to each of 3 friends—Aban, Hasan, and Kamil. Find the chance that Aban gets precisely 3 reds, Hasan gets exactly 7 reds, and Kamil gets 2 reds.

Solution:
The number of ways to deal the cards is $\binom{24}{8,\,8,\,8} = \frac{24!}{8!^3} = 9{,}465{,}511{,}770$

The number of ways to deal such that the friends get the specified number of reds is:
$$\binom{12}{3,\,7,\,2} * \binom{12}{5,\,1,\,6} = 7{,}920 * 5{,}544 = 43{,}908{,}408$$

Assembling the final solution produces:
$$\Pr = \frac{\binom{12}{3,\,7,\,2} * \binom{12}{5,\,1,\,6}}{\binom{24}{8,\,8,\,8}} = \frac{43{,}908{,}408}{9{,}465{,}511{,}770} = 0.004639$$

Notice that all the top portions of the multinomial coefficients sum to 24. In addition, all the bottom portions sum to 24. All cards are accounted for. In addition, within each multinomial coefficient, the individual group sizes in the bottom sum to the full pool size in the top.

Chapter 3: Basics of Probability Distributions

A probability distribution describes the relative or actual probability that a random variable can assume each real value. Random variables form a counterpart to events. The random variables in this book can take on just real numbers and not qualitative values. A random variable could assume infinitely many values, or perhaps it takes a small, finite set of integers. Correspondingly, random variables can be continuous, discrete, or a mixture of continuous intervals and small sets of values. Every probability distribution needs to follow the three fundamental probability axioms. A probability distribution could follow a template distribution, or it could have a unique, one-of-a-kind form. This chapter introduces two closely connected distributions—the Bernoulli and binomial distributions—as examples of basic discrete probability distributions. The continuous uniform distribution is also provided for random variables which can only assume values on an interval with equal probability.

Section 3-1: Random Variables

A *random variable* is a function on the sample space. The random variable takes an outcome from the sample space as input and will output a real value. A random variable is written as X(s). However, often just the capital letter, such as X, denotes the random variable. A specific value of the random variable is denoted by a lowercase letter, such as x.

As an example of a random variable, suppose that Uriah is performing an experiment in which he flips a fair coin 4 times. Uriah writes the outcome of each experiment as an ordered set of coin outcomes, such as {H, T, H, H}. However, Uriah is really just interested in how many heads he obtained in the experiment. He can define the random variable X(s) to take the current outcome from the experiment and count the quantity of heads. The value of X(s) will then be an integer between 0 and 4, inclusive.

Random variables could be the result of arithmetic applied to the outcome. If the experiment's outcome is a real value, then the random variable could be the outcome squared, cubed, or multiplied by some coefficient.

The word *random* has a precise meaning in probability. A random variable is assigned a value in an unbiased and unhampered fashion. The person performing the experiment cannot give favor to any outcome. A random process does not give unfair advantage to a particular value which could be chosen. Randomness does not mean that every value has the same chance of being chosen. In most distributions (excluding a uniform distribution), some values will be more likely than others. A random selection means that a value has a chance of being chosen which is proportional to its relative likelihood among other potential values.

Random variables can take on an enormous variety of values. Some may be limited to a certain range, while others extend to positive and negative infinity. Variables can be integer-valued, floating-point-valued, or take on English characters that do not translate easily to numbers. For example, a random variable could be the color produced in some experiment. Probability is typically concerned with just quantitative random variables rather than qualitative random variables. That is, a random variable as studied in probability usually has a numerical value.

A random variable can be discrete, continuous, or mixed. A *discrete random variable* has at most countably many unique values. A *continuous random variable* has uncountably many different values. Continuous random variables are usually defined over one or several continuous intervals (including, perhaps, R). A *mixed random variable* has some continuous intervals of values along with discrete, individual values.

A *random sample* consists of n outcomes from an experiment. That is, the experiment is run n times, producing an outcome after every trial. Each outcome is modeled as a random variable. Typically, the ith outcome is denoted as X_i, so that the outcomes are $X_1, X_2, \ldots, X_n$. All the X_i's are iid; each trial must be run in the same fashion. The *sample size* is the number of trials, n.

Section 3-2: Structure of Discrete Probability Distributions

A discrete random variable X has a *probability function* which assigns positive probability to at most countably many x-values. Typically, X has a finite quantity of potential values, such as a subset of several integers. The probability function, often abbreviated "pf," reveals the probability that X = x at each x-value. The function f(x) denotes the probability function. When graphed, a pf will not be a continuous curve; the pf will consist of a series of points. You can represent a pf in table form or with a piecewise function. It is good practice to explicitly specify zero probability at impossible x-values, especially when the pf is written as a piecewise function.

The probability function must follow the basic axioms of probability. In accordance with the first axiom, each value of f(x) must be between 0 and 1, inclusive. Additionally, the second axiom requires that all the f(x)'s sum to 1.

Example

A video rental store allows customers to rent movies for up to 5 days. After the fifth day, the customer must buy the movie. The number of days, X, that a customer has rented a movie is rounded up to the nearest integer. In other words, an actual outcome could be 2.6 days, but the random variable computes X(s) = ceil(s) and would produce 3.

The store owner computes the following probability function:

x	f(x)
1	0.08
2	0.14
3	0.35
4	0.17
5	0.12
6+	0.14

Notice that each probability value is between 0 and 1, inclusive. All the probabilities sum to 1. The table format implicitly assumes that any unmentioned x-values have probability zero. For instance, $\Pr(X = -5) = 0$. In addition, the range of x values for integers greater than or equal to 6 have been grouped into the "6+" category.

For this problem, the probability function should not be written in equation format since each x-value has a unique probability, and there is no simple way to create a discrete equation which produces the specified probabilities for all the x-values.

A plot of the probability function appears here:

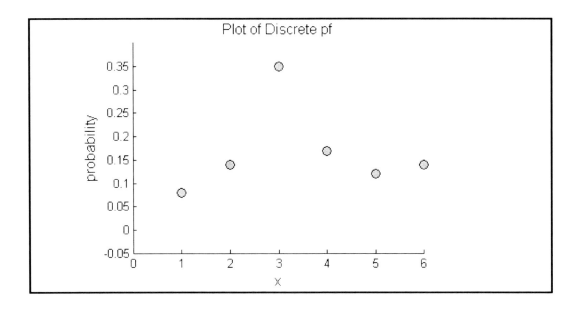

When graphing a probability function, you should place a solid dot at each ordered pair $(x, f(x))$. You could also plot a bar reaching up from the x-axis to the point. Technically, a sequence of open circles and solid segments should run along the x-axis between any two points, since the probabilities of these x-values are zero. However, the graph is usually most clear by showing just the points with positive probability.

Section 3-3: Structure of Continuous Probability Distributions

A continuous random variable has a *probability density function* (or pdf) showing the relative chance that X takes each potential value. A pdf is denoted by the function f(x). A pdf can generate values between 0, inclusive, and positive infinity, exclusive. A pdf cannot be negative, however. If f(a) = 1 and f(b) = 2, then b is twice as likely to occur as a. The pdf can approach positive infinity at a vertical asymptote.

A counterintuitive fact of a continuous random variable is that $\Pr(X = x) = 0 \quad \forall \, x$. In words, the mathematical chance that the variable takes any of its potential values is zero. However, it does not mean that it is impossible for X = x. If it was impossible for X to take any of its values, then Pr(S) = 0, which violates the second probability axiom. The probability at any x-value must be zero because if the probability was positive for infinitely many x-values, then Pr(S) would be infinity. The value of the pdf f(x) just shows the relative (rather than actual) probability of X taking on x.

The integral under a pdf over all x-values must be 1. This rule satisfies the second probability axiom.

Example
The probability density function for a random variable X is:
$$f(x) = \begin{cases} 0.3e^{-0.3x}, & \text{if } x > 0 \\ 0, & \text{else} \end{cases}$$

The probability density function is graphed below:

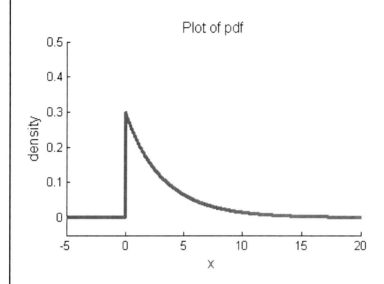

Let's confirm that f is truly a pdf.
Check #1: All the values of f must be greater than or equal to 0.
    $f(x) = 0$ for $x \leq 0$.
    f(x) is strictly positive when x > 0.
Check #2: The integral of f over all real values must be 1.
$$\int_0^\infty 0.3e^{-0.3x}\,dx = \left[0.3 * \frac{1}{-0.3} e^{-0.3x}\right]_0^\infty = -1(e^{-\infty} - e^0) = -1(0-1) = 1$$

Example
The pdf for a random variable X is:
$$f(x) = \begin{cases} cx^2, & \text{for } -2 \leq x \leq 3 \\ 0, & \text{else} \end{cases}$$

Let's solve for c which allows f to be a pdf.

$$\int_{-2}^{3} cx^2\,dx = c * \left[\frac{1}{3}x^3\right]_{-2}^{3} = c * \frac{1}{3} * (3^3 - (-2)^3) = c * \frac{1}{3} * (27 - (-8))$$

$$c * \frac{1}{3} * 35 = 1 \rightarrow c * \frac{35}{3} = 1 \rightarrow c = 3/35$$

The pdf now becomes

$$f(x) = \begin{cases} \frac{3}{35} x^2, & \text{for } -2 \leq x \leq 3 \\ 0, & \text{else} \end{cases}$$

The pdf's graph is:

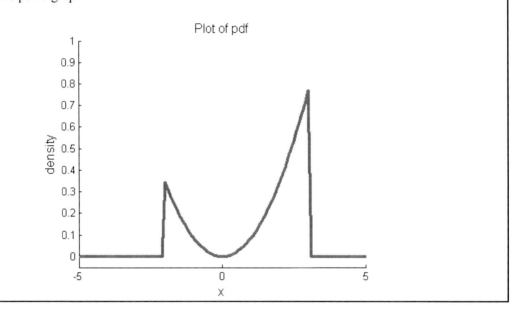

Example

An unusual gambling game requires the player to spin a three-sided top with the colors yellow, magenta, and cyan. The top is fair and balanced. Depending on the top color, the player receives a payoff randomly selected from an interval. The intervals are:

yellow: [-5, 9]
magenta: [-2, 4]
cyan: [-1, 7]

(a.) Find the chance that the player receives a payoff which is positive.
(b.) Find the chance that the player's value is within the interval (-3, -1).

Solutions:
The color of the top is discrete, while the actual payoff is selected from a continuous interval.

(a.) The chance of each top color is 1/3.

> Pr(positive payoff) = Pr(yellow AND payoff in (0, 9]) +
> Pr(magenta AND payoff in (0, 4]) + Pr(cyan AND payoff in (0, 7])
>
> Pr(yellow AND payoff in (0, 9]) = Pr(yellow) * Pr(payoff in (0, 9] | yellow)
> $$= \frac{1}{3} * \frac{9}{14} = \frac{9}{42}$$
>
> Pr(magenta AND payoff in (0, 4]) = Pr(magenta) * Pr(payoff in (0, 4] | magenta)
> $$= \frac{1}{3} * \frac{4}{6} = \frac{4}{18}$$
>
> Pr(cyan AND payoff in (0, 7]) = Pr(cyan) * Pr(payoff in (0, 7] | cyan)
> $$= \frac{1}{3} * \frac{7}{8} = \frac{7}{24}$$
>
> Pr(positive payoff) = $\frac{9}{42} + \frac{4}{18} + \frac{7}{24} = \frac{367}{504} \approx 0.72817$

A distribution can also be mixed, in which case the variable could take on a value within one of several intervals or a value from a discrete set. For a mixed distribution, the function f(x) is called the "mixed pf/pdf." The combined sum of integrating f over continuous regions of x and summing f over discrete values of x must be 1.

> **Example**
> The random variable X can take the values 0, 1, or 2 with probabilities 0.12, 0.24, and 0.34, respectively. For values of X in the interval (3, 4), the probability density function is $f(x) = 1 - 0.2x$.
> (a.) What is the probability that X lies within [2, 3.5]?
> (b.) What is the chance that X is below 3.25?
>
> Solutions:
>
> (a.) $\Pr(X \in [2, 3.5]) = f(2) + \int_{3}^{3.5} 1 - 0.2x \, dx = 0.34 + \left[x - \frac{0.2}{2} x^2 \right]_{3}^{3.5}$
> $$= 0.34 + 0.175 = 0.515$$
>
> (b.) $\Pr(X < 3.25) = f(0) + f(1) + f(2) + \int_{3}^{3.25} 1 - 0.2x \, dx$
> $$= 0.12 + 0.24 + 0.34 + \left[x - \frac{0.2}{2} x^2 \right]_{3}^{3.25} = 0.79375$$

Section 3-4: Distribution Functions

The *distribution function* for a random variable X measures the variable's cumulative probability. The distribution function is denoted with a capital letter, such as F. An abbreviation for the distribution function is "df." F measures the probability that X is less than or equal to a particular x-value: $F(x) = \Pr(X \leq x)$. The df is cumulative in the sense that it sums all the probabilities for x-values at or below X = x. The df must exist at all $x \in R$. Output values of F(x) must be between 0 and 1, inclusive.

The distribution function is a non-decreasing function. F(x) can either increase or remain constant, but it cannot decrease. The facts that F(x) must be between 0 and 1 and it is non-decreasing mean that F(x) must be around 0 at the lowest possible x-value and must be around 1 at the highest possible x-value.

Limits of the Distribution Function

$$\lim_{x \to -\infty} F(x) = 0 \qquad \lim_{x \to \infty} F(x) = 1$$

Mathematicians discriminate between evaluating F(x) as x-values approach x from the left and from the right. $F(x^+)$ denotes the value of F as X approaches x from the right (i.e., as x-values which are slightly larger than x decrease and come closer to x). $F(x^-)$ signifies the value of F as X approaches x from the left (x-values that are slightly smaller than x increase towards x). Distribution functions are right continuous, requiring that $F(x) = F(x^+)$.

The plot below shows the distribution function for a discrete random variable. Consider the values of F(x) around x = 6. When approaching x = 6 from the left, the distribution function yields 0.3 (that is, $F(x^-) = 0.3$). However, approaching x = 6 from the right produces the value $F(x^+) = 0.5$. Since F is right continuous, $F(x) = F(x^+) = 0.5$

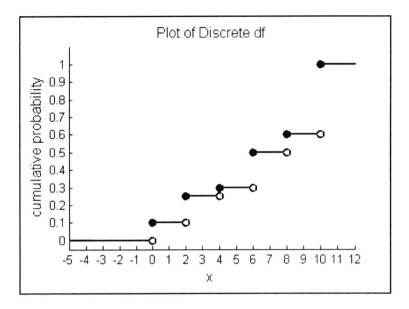

The probability that X = x is equivalent to the magnitude of the jump in the distribution function at x. In other words, $\Pr(X = x) = F(x^+) - F(x^-)$. If the distribution function is continuous at x and makes no jump, then $\Pr(X = x) = 0$. In the previous example, the probability that X takes 6 is $F(x^+) - F(x^-) = 0.2$.

The survival function, denoted as s(x), is the complement of the distribution function. s(x) gives the chance that X survives past x. That is, $s(x) = \Pr(X > x)$. Note that s(x) and F(x) are related in $s(x) + F(x) = 1$. The survival function is used heavily in the development of life insurance and long-term care insurance, as analysts are concerned about the chance that people live beyond a given age.

Distribution Functions for Discrete Random Variables

When X is a discrete random variable, its distribution function is a step function. F will consist of segments, each of which has a closed circle on the left side and an open circle on the right side. The fact that the distribution function is right continuous requires the closed circle on the left endpoint of each segment. The distribution function is constant on each step because X cannot assume any values over the corresponding x-interval. If you are given the distribution function and must deduce the probability function at X = x, you can compute the gap between the step to the left of x and the step to the right of x and assign the difference to f(x).

Development of a Distribution Function from a pf
$$F(x) = \sum_{u \ni f(u) > 0, u \leq x} f(u)$$

Digital Actuarial Resources   Comprehensive Probability Review for Actuarial Exams

Example

Construct the distribution function for a discrete random variable with this probability function:

x	f(x)
2	0.28
5	0.16
6	0.31
8	0.25

Solution:
A pre-processing step before graphing F(x) is to build a new probability table. You should include a column for F(x). In addition, the x-values are now ranges of x-values.

x	F(x)
$x < 2$	0
$2 \leq x < 5$	0.28
$5 \leq x < 6$	0.44
$6 \leq x < 8$	0.75
$x \geq 8$	1.0

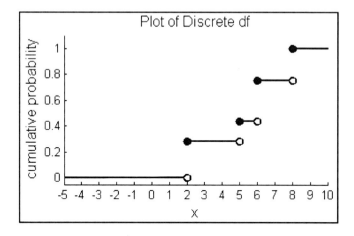

Example

You are faced with the distribution function provided below:

x	F(x)
$x < -3$	0
$-3 \leq x < 0.5$	0.19
$0.5 \leq x < 7.5$	0.79

$7.5 \leq x < 10$	0.81
$10 \leq x < 11$	0.85
$x \geq 11$	1.0

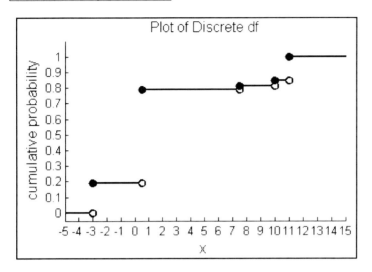

Find the probability function and graph it.

Solution:
Compute the gap between each pair of steps and build a table.

x	f(x)
-3	0.19
0.5	0.60
7.5	0.02
10	0.04
11	0.15

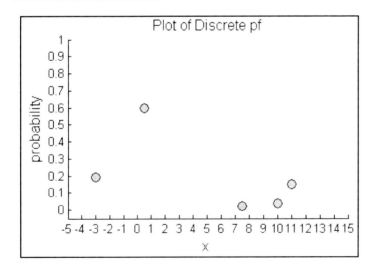

Distribution Functions for Continuous Random Variables

If X is a continuous random variable, then the distribution function is found by integrating the probability density function. Integrating in the continuous case corresponds to summing in the discrete case. The area under the pdf from $-\infty$ to x represents the chance that X assumes any value within $(-\infty, x)$. As long as the pdf is non-negative, the distribution function will always increase or remain constant (the area under the pdf will be non-negative).

Development of a Distribution Function from a pdf

$$F(x) = \int_{-\infty}^{x} f(u)\, du$$

We must replace x with a dummy variable, such as u, within the integral because x will be the upper bound.

The computation of the distribution function for a continuous random variable is illustrated in the plot below. The example pdf is a fourth-degree polynomial. To compute $F(X = 6)$, we need to determine the area of the shaded region. The shaded part represents the cumulative probability from the smallest possible x-value to $X = 6$.

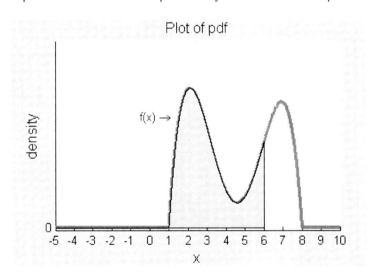

Example
Find the distribution function for a random variable X with this pdf:

$$f(x) = \begin{cases} \dfrac{20}{139}(x^4 + 0.2x^3), & \text{if } 1 < x < 2 \\ 0, & \text{else} \end{cases}$$

Solution:
When computing F(x), the lower bound of the integral should be the smallest x-value for which the current piece of the pdf applies. In this case, the pdf is 0 when $x \leq 1$, so there is no need to integrate the pdf from $-\infty$ to 1.

$$F(x) = \int_1^x \frac{20}{139}(u^4 + 0.2u^3)\, du = \frac{20}{139}\left[\frac{1}{5}u^5 + \frac{0.2}{4}u^4\right]_1^x$$

$$= \frac{20}{139}\left(\left(\frac{1}{5}x^5 + \frac{0.2}{4}x^4\right) - \left(\frac{1}{5}1^5 + \frac{0.2}{4}1^4\right)\right)$$

$$= \frac{20}{139}\left(\left(\frac{1}{5}x^5 + \frac{0.2}{4}x^4\right) - 0.25\right) = \frac{4}{139}x^5 + \frac{1}{139}x^4 - \frac{5}{139}, \text{ when } 1 < x < 2$$

$$F(x) = \begin{cases} 0, & \text{if } x \leq 1 \\ \dfrac{4}{139}x^5 + \dfrac{1}{139}x^4 - \dfrac{5}{139}, & \text{if } 1 < x < 2 \\ 1, & \text{if } x \geq 2 \end{cases}$$

Most of the distribution functions for continuous random variables are written as piecewise functions. Typically, F(x) = 0 for x-values below some minimum x-value, and F(x) = 1 when x surpasses some maximum x-value.

Example
Find the df for the random variable X with this mixed pf/pdf:

$$f(x) = \begin{cases} 0.3, & \text{if } x = 1 \\ -0.5x^2 + 3x - 4, & \text{if } 2 < x < 4 \\ 1/30, & \text{if } x = 6 \\ 0, & \text{else} \end{cases}$$

Solution:
Let's first determine the df for 2 < x < 4

$$F(x) = \int_2^x -0.5u^2 + 3u - 4 \, du = \left[\frac{-0.5}{3}u^3 + \frac{3}{2}u^2 - 4u \right]_2^x$$

$$= \left(\frac{-0.5}{3}x^3 + \frac{3}{2}x^2 - 4x \right) - \left(\frac{-0.5}{3}2^3 + \frac{3}{2}2^2 - 4*2 \right)$$

$$= \frac{-0.5}{3}x^3 + \frac{3}{2}x^2 - 4x + \frac{10}{3}$$

For 2 < x < 4, we must also add-on all the earlier probabilities. The earlier probability occurs at x = 1. In the end, when 2 < x < 4,

$$F(x) = \frac{-0.5}{3}x^3 + \frac{3}{2}x^2 - 4x + \frac{10}{3} + 0.3 = -\frac{1}{6}x^3 + \frac{3}{2}x^2 - 4x + \frac{109}{30}$$

Overall,

$$F(x) = \begin{cases} 0, & \text{if } x < 1 \\ 0.3, & \text{if } 1 \leq x \leq 2 \\ -\frac{1}{6}x^3 + \frac{3}{2}x^2 - 4x + \frac{109}{30}, & \text{if } 2 < x < 4 \\ 29/30, & \text{if } 4 \leq x < 6 \\ 1, & \text{if } x \geq 6 \end{cases}$$

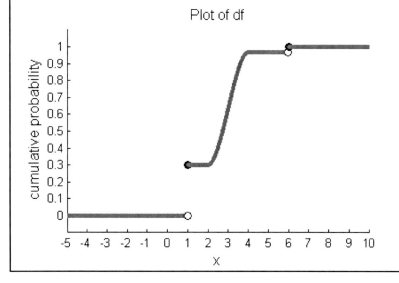

You can differentiate the distribution function to produce the probability density function for a continuous random variable: $\frac{dF}{dx} = f(x)$.

Example

You are given the distribution function $F(x) = \begin{cases} 0, & \text{if } x \leq 0 \\ 1 - e^{-0.6x}, & \text{if } x > 0 \end{cases}$

Find the pdf for X.

Solution:
Since the df is continuous for x > 0, we can differentiate the df over this interval.

$$\frac{dF}{dx} = \frac{d}{dx}\left[1 - e^{-0.6x}\right] = 0 - e^{-0.6x} * (-0.6) = 0.6e^{-0.6x}$$

$f(x) = \begin{cases} 0, & \text{if } x \leq 0 \\ 0.6e^{-0.6x}, & \text{if } x > 0 \end{cases}$

If X is continuous over the interval (a, b), then you can find the probability that X lies between a and b by integrating the pdf over the interval: $\Pr(a < X < b) = \int_a^b f(x)\, dx$.

The fact that the probability includes or excludes a and b has no bearing on the bounds of integration. Recall that the probability of X = x for a continuous random variable must be zero. As a consequence, deciding to include or exclude the bounds (single points) will not affect the probability for the interval. In summary, for a continuous random variable, $\Pr(a < X < b) = \Pr(a \leq X \leq b) = \Pr(a \leq X < b)$ and so on. If you already have the df, then you can compute $\Pr(a < X < b) = \Pr(X < b) - \Pr(X \leq a) = F(b) - F(a)$.

Example

The pdf for the random variable X is approximately:

$f(x) = \begin{cases} -0.01752x^2 + 0.1401x - 0.06566, & \text{if } 0.5 < x < 7.5 \\ 0, & \text{else} \end{cases}$

(a.) Compute $\Pr(4 \leq X \leq 6)$ without finding the distribution function.
(b.) Find the distribution function and compute $\Pr(3 < X < 4)$.

Solutions:
(a.)
$$\int_4^6 -0.01752x^2 + 0.1401x - 0.06566\, dx = \left[\frac{-0.01752}{3}x^3 + \frac{0.1401}{2}x^2 - 0.06566x\right]_4^6$$

$$= \left(\frac{-0.01752}{3}6^3 + \frac{0.1401}{2}6^2 - 0.06566*6\right) - \left(\frac{-0.01752}{3}4^3 + \frac{0.1401}{2}4^2 - 0.06566*4\right)$$

$$= 0.8664 - 0.4844 = 0.382$$

(b.)

When $0.5 < x < 7.5$,

$$F(x) = \int_{0.5}^{x} -0.01752u^2 + 0.1401u - 0.06566 \ du$$

$$= \left[\frac{-0.01752}{3}u^3 + \frac{0.1401}{2}u^2 - 0.06566u \right]_{0.5}^{x}$$

$$= \left(\frac{-0.01752}{3}u^3 + \frac{0.1401}{2}u^2 - 0.06566u \right) - \left(\frac{-0.01752}{3}0.5^3 + \frac{0.1401}{2}0.5^2 - 0.06566*0.5 \right)$$

$$= \frac{-0.01752}{3}u^3 + \frac{0.1401}{2}u^2 - 0.06566u + 0.01605$$

$\Pr(3 < X < 4) = F(4) - F(3) = 0.50045 - 0.29184 = 0.20861$

Section 3-5: The Quantile Function

The *quantile function* is the inverse of the distribution function. The quantile function produces x-values at which probabilities occur. The notation for the quantile function appears here:

$F^{-1}(p) = x$

The probability values that you input into the function must be between 0 and 1, exclusive. You cannot provide $p = 0$ or $p = 1$ because some distribution functions never actually produce these probabilities; the distribution function could approach the probabilities as horizontal asymptotes. The output value from $F^{-1}(p)$ is called the p quantile or the $100*p^{th}$ percentile; $100*p\%$ of the data lie at or below the corresponding x-value.

Example
What is the quantile function for this distribution function?:

$$F(x) = \begin{cases} 0, & \text{if } x < 2.60546 \\ -0.07x^2 + 0.95x - 2, & \text{if } 2.60546 \leq x \leq 5 \\ 1, & \text{if } x > 5 \end{cases}$$

In addition, find the 15th percentile and the 90th percentile.

Solution:
Invert the distribution function over the important interval.

$p = -0.07x^2 + 0.95x - 2$
$0 = -0.07x^2 + 0.95x - 2 - p$

$$x = \frac{-0.95 \pm \sqrt{0.95^2 - 4*(-0.07)*(-2-p)}}{2*(-0.07)} = \frac{-0.95 + \sqrt{0.9025 + 0.28*(-2-p)}}{-0.14}$$

The quantile function is $F^{-1}(p) = \dfrac{-0.95 + \sqrt{0.9025 + 0.28*(-2-p)}}{-0.14}$ for $p \in (0, 1)$

15th percentile = $F^{-1}(0.15) = 2.87015$

90th percentile = $F^{-1}(0.90) = 4.6369$

When X is discrete, it is possible that an x-value does not exist which directly satisfies F(x) = p. In this case, when finding the p quantile, choose the lowest possible x-value at which the cumulative probability p is just met or exceeded.

Example
The random variable X has this pf:

$$f(x) = \begin{cases} 0.11, & \text{if } x = 0 \\ 0.35, & \text{if } x = 1 \\ 0.28, & \text{if } x = 2 \\ 0.05, & \text{if } x = 3 \\ 0.21, & \text{if } x = 4 \\ 0, & \text{else} \end{cases}$$

Evaluate these quantiles: (a.) 50th (b.) 1st (c.) 95th

Solutions:
First, let's build the distribution function in table form.

x	F(x)
X < 0	0
0 ≤ X < 1	0.11
1 ≤ X < 2	0.46
2 ≤ X < 3	0.74
3 ≤ X < 4	0.79
X ≥ 4	1

(a.) What is the smallest x-value with a corresponding df-value at or above 0.5?
$F^{-1}(0.5) = 2$

(b.) $F^{-1}(0.01) = 0$
(c.) $F^{-1}(0.95) = 4$

Quartiles

Every probability distribution has *quartiles* which break the probability into fourths. Quartiles are particular quantiles of interest to analysts. An x-value exists at each quartile. The *lower quartile* is the 25th percentile. 25% of data should lie at or below the lower quartile. The *upper quartile* is the 75th percentile. The *median* is the 50th percentile. All probability distributions must have lower and upper quartiles, along with a median. The difference between the upper quartile and lower quartile is called the *inter-quartile range*.

The quartiles are illustrated in the *box-and-whiskers plot* below. The left and right sides of the box delimit the lower and upper quartiles, respectively. The median is marked as a vertical segment through the box. Whiskers then extend to the minimum and maximum values discovered in an experimental data set.

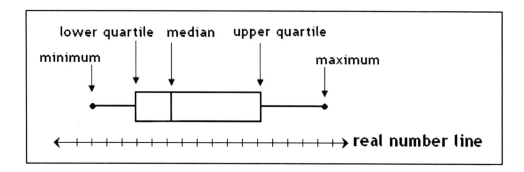

Example
Determine the 25th, 50th, and 75th percentiles for a distribution with this pf:

$$f(x) = \begin{cases} 0.37, & \text{if } x = 5 \\ 0.12, & \text{if } x = 10 \\ 0.04, & \text{if } x = 15 \\ 0.21, & \text{if } x = 20 \\ 0.19, & \text{if } x = 25 \\ 0.07, & \text{if } x = 30 \\ 0, & \text{else} \end{cases}$$

Solutions:

Build the distribution function in tabular form.

x	F(x)
X < 5	0
5 ≤ X < 10	0.37
10 ≤ X < 15	0.49
15 ≤ X < 20	0.53
20 ≤ X < 25	0.74
25 ≤ X < 30	0.93
X ≥ 30	1

25th percentile = (X = 5)
50th percentile = (X = 15)
75th percentile = (X = 25)

Example
A random variable X has this pdf:
$$f(x) = \begin{cases} 3.5e^{-3.5x}, & \text{if } x > 0 \\ 0, & \text{else} \end{cases}$$

Find the 25th, 50th, and 75th percentiles.

Solutions:
First, find the distribution function.

$$F(x) = \int_0^x 3.5e^{-3.5u}\, du = \left[-e^{-3.5u}\right]_0^x = -e^{-3.5x} + e^{-3.5*0} = 1 - e^{-3.5x}, \text{ if } x > 0$$

Second, invert the distribution function to build the quantile function.

$$p = 1 - e^{-3.5x} \rightarrow p - 1 = -e^{-3.5x} \rightarrow 1 - p = e^{-3.5x} \rightarrow \ln(1-p) = -3.5x$$

$$\frac{\ln(1-p)}{-3.5} = x = F^{-1}(p)$$

25th percentile:
$$F^{-1}(0.25) = \frac{\ln(1-0.25)}{-3.5} = 0.08219 = x$$

50th percentile:
$$F^{-1}(0.50) = \frac{\ln(1-0.50)}{-3.5} = 0.19804 = x$$

75th percentile:
$$F^{-1}(0.75) = \frac{\ln(1-0.75)}{-3.5} = 0.39608 = x$$

Section 3-6: Bernoulli Distribution

The Bernoulli distribution is perhaps the simplest probability distribution with a formal name. The Bernoulli distribution is discrete. The random variable X with a Bernoulli distribution can take one of two values: 0 or 1. The value '0' often corresponds to failure, while '1' means success. The Bernoulli distribution is ideal for modeling a trial which can either fail or succeed. The distribution has one parameter, which is p, representing the probability that X = 1. The complement of p is denoted by q, which represents the chance that X = 0.

Notation for the Bernoulli Distribution
X = Boolean status of trial
p = probability of success

The Bernoulli probability function can be represented in tabular or equation form. The tabular form is:

x	f(x)
0	1 − p
1	p

The equation form is:

$$f(x) = \begin{cases} p^x(1-p)^{1-x}, & \text{if } x = 0 \text{ or } 1 \\ 0, & \text{else} \end{cases}$$

When graphed, the Bernoulli pf has just two significant points at (0, 1 − p) and (1, p).

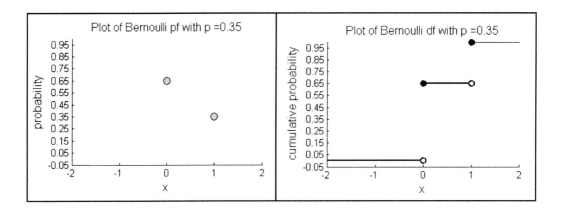

The Bernoulli distribution function is a simple step function with three total steps. The df is 0 for X < 0. A jump of magnitude (1 – p) occurs at X = 0. An open circle exists at (0, 0) and a closed circle exists at (0, 1 – p), illustrating the discontinuity. The df then remains level until X = 1, at which point the df jumps up to 1. Upon reaching X = 1, the df has covered all the x-values with positive probability and cannot increase any further, so the df remains level at 1. The df in equation form is:

$$F(x) = \begin{cases} 0, & \text{if } x < 0 \\ (1-p), & \text{if } 0 \leq x < 1 \\ 1, & \text{if } x \geq 1 \end{cases}$$

Example
Give the pf and df for a Bernoulli distributed random variable X with…
(a.) p = 0.15
(b.) p = 0.42

Solutions:
(a.)

x	f(x)
0	0.85
1	0.15

$$F(x) = \begin{cases} 0, & \text{if } x < 0 \\ 0.85, & \text{if } 0 \leq x < 1 \\ 1, & \text{if } x \geq 1 \end{cases}$$

(b.)

x	f(x)
0	0.58
1	0.42

$$F(x) = \begin{cases} 0, & \text{if } x < 0 \\ 0.58, & \text{if } 0 \leq x < 1 \\ 1, & \text{if } x \geq 1 \end{cases}$$

Example
Plot the pf and df for the following random variables:
(a.) $X \sim \text{Bernoulli}(0.25)$
(b.) $X \sim \text{Bernoulli}(0.78)$

Solutions:
(a.)

(b.)

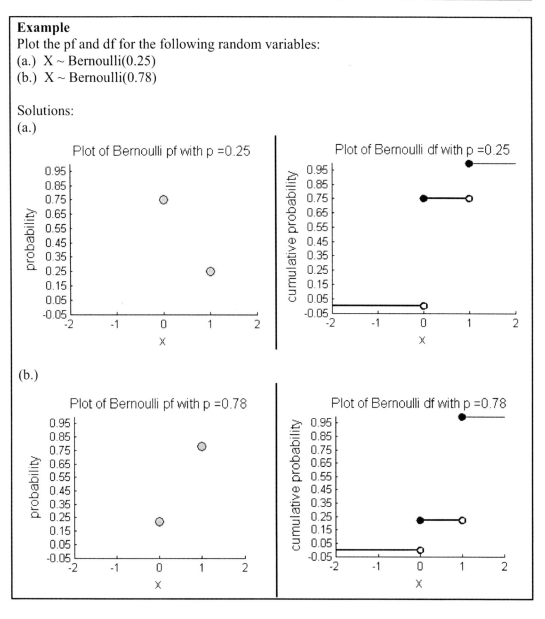

Section 3-7: Binomial Distribution

A binomial experiment consists of a finite sequence of Bernoulli trials. All the trials are identical and presumably occur in sequential order. Each trial has two potential results—success and failure (hence the term "binomial"). Only one outcome is possible in each trial. The letter 'p' signifies the probability of success, while 'q' denotes the probability of failure. The potential outcomes are complementary so that $p + q = 1$. The probabilities of success and failure remain constant over the trials. The ith trial can be represented as the random variable $X_i \sim \text{Bernoulli}(p)$. In addition, the trials are independent of each other. Whatever happened on the previous trials does not affect the chance of success on the next trial. For instance, if an experiment with a fair coin produced 100 tails on the first 100 trials, the chance of heads on the 101st trial is still 50%. We cannot assume that the experimental proportion of successes implies a different p-value as used in the original distribution. The binomial distribution has two parameters: p and n. The value n is the quantity of trials.

The random variable X with a binomial distribution stores the number of successes obtained during the trials. Since each trial has a Bernoulli distribution and produces either $X = 0$ or $X = 1$, you can simply sum the results of all Bernoulli runs to find the total number of successes. That is, $X = X_1 + X_2 + ... + X_n$. X could be any integer from 0 to n, inclusive. It is possible that all trials failed or all trials succeeded. Since X could assume just a finite number of different values, a binomially distributed random variable is discrete.

Notation for the Binomial Distribution
X = # of successes
p = probability of success on a trial
n = # of trials

The binomial pf is revealed and discussed below. The pf is positive only when the random variable X is an integer between 0 and n, inclusive. The three main components of the pf include a combination, a factor for the successes, and a factor for the failures.

Binomial Probability Function

$$f(x) = \begin{cases} \binom{n}{x} \cdot p^x \cdot q^{n-x}, & \text{for x an integer between 0 and n, inclusive} \\ 0, & \text{else} \end{cases}$$

The most fundamental use of a binomial distribution is to determine the chance of k successes within a total of n trials. Since the events are independent, we can use the equation $\Pr(A \text{ AND } B) = P(A) * P(B)$. If we consider just the successes, we are interested in:

$$\underbrace{P(\text{success AND success AND success} \ldots)}_{k \text{ times}} = \underbrace{P(\text{success}) * P(\text{success}) * P(\text{success})\ldots}_{k \text{ times}}$$

$$= p * p * p \ldots = p^k$$

Thus, the probability of obtaining k successes is p^k. The cases that did not succeed must have failed, so (n – k) failures exist. The probability of (n – k) failures is q^{n-k}.

The binomial pf features a combination expression because there are multiple ways to select x successes from among all n trials. For example, the first x trials could all be successes, followed by (n – x) failures. Or, all x successes could come at the end of the experiment. Or, a success could occur, followed by a few failures, followed by another success, and so on. In other words, there are many ways to obtain exactly x successes and (n – x) failures. We need to express the number of ways to order x successes within a pool of n total trials, which is accomplished with ${}_nC_x$.

The distribution function for a binomially distributed random variable is a step function. The df is level at zero for x-values below 0. At every x-value between 0 and n, inclusive, a jump occurs equivalent to $\Pr(X = x)$. After X reaches n, the df levels off at 1. A few pf's and df's for binomial distributions appear below:

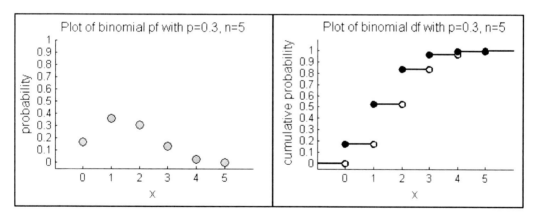

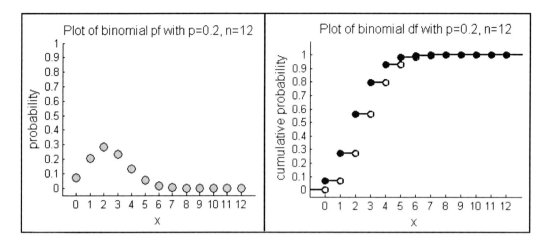

To compute the probability that X lies in some interval, you can simply sum f(x) for all x-values in that interval. Summing individual probabilities is legal because all the x-values are disjoint. It is impossible for a binomial experiment to produce exactly 5 successes and exactly 10 successes in the same run. You need to be careful about including or excluding the interval endpoints.

It may also be more convenient to compute the probability of the complement of the event of interest and then subtract the probability value from 1. For instance, suppose n = 100, and you need to find the probability that X exceeds 3. One option is to calculate $f(4) + f(5) + ... + f(100)$. However, this route becomes tedious and error prone very fast. The more sound strategy is to compute $f(0) + f(1) + f(2) + f(3)$ and then subtract the sum from 1. The sets of x-values {0, 1, 2, 3} and {4, 5, ..., 100} are complements of each other.

Example
A golfer is putting on a green. He hits 10 balls from the same location. The chance of making the hole is 34%.
(a.) What is the probability the golfer makes exactly 6 shots?
(b.) ...exactly 2 shots?
(c.) ...at least 7 shots?
(d.) ...between 3 and 5 shots, inclusive?

Solutions:
Let X be the variable signifying the number of shots made out of the 10 possible.
(a.) $P(X = 6) = {}_{10}C_6 * 0.34^6 * 0.66^4 = 210 * 0.0015448 * 0.18975 = 0.061556$
(b.) $P(X = 2) = {}_{10}C_2 * 0.34^2 * 0.66^8 = 45 * 0.1156 * 0.036 = 0.18729$

(c.) P(at least 7 successes) = P(7 or 8 or 9 or 10 successes) = $P(X \geq 7)$
= $P(X = 7) + P(X = 8) + P(X = 9) + P(X = 10)$
= ${}_{10}C_7 * 0.34^7 * 0.66^3 + {}_{10}C_8 * 0.34^8 * 0.66^2 + {}_{10}C_9 * 0.34^9 * 0.66^1 + {}_{10}C_{10} * 0.34^{10} * 0.66^0$

$$= 120 * 0.000525 * 0.2875 + 45 * 0.000179 * 0.4356 + 10 * 0 * 0.66 + 1 * 0 * 1$$
$$= 0.01811 + 0.00351 + 0 + 0$$
$$= 0.02162$$

(d.) P(between 3 and 5 successes) = $P(3 \leq X \leq 5) = P(X = 3) + P(X = 4) + P(X = 5)$
$$= {}_{10}C_3 * 0.34^3 * 0.66^7 + {}_{10}C_4 * 0.34^4 * 0.66^6 + {}_{10}C_5 * 0.34^5 * 0.66^5$$
$$= 120 * 0.0393 * 0.05455 + 210 * 0.01336 * 0.08265 + 252 * 0.00454 * 0.12523$$
$$= 0.25726 + 0.23188 + 0.14327$$
$$= 0.63241$$

Example
Build the probability distribution for a binomial experiment with n = 5 and q = 0.27.

Solution
p = 1 − q = 0.73

We need the value of P(X) when X spans 0 to 5, inclusive.

$P(X = 0) = {}_5C_0 * 0.73^0 * 0.27^5 = 1 * 1 * 0.00143 = 0.00143$
$P(X = 1) = {}_5C_1 * 0.73^1 * 0.27^4 = 5 * 0.73 * 0.0053144 = 0.0194$
$P(X = 2) = {}_5C_2 * 0.73^2 * 0.27^3 = 10 * 0.5329 * 0.01968 = 0.10489$
$P(X = 3) = {}_5C_3 * 0.73^3 * 0.27^2 = 10 * 0.38901 * 0.0729 = 0.2836$
$P(X = 4) = {}_5C_4 * 0.73^4 * 0.27^1 = 5 * 0.28398 * 0.27 = 0.38338$
$P(X = 5) = {}_5C_5 * 0.73^5 * 0.27^0 = 1 * 0.2073 * 1 = 0.2073$

x	P(x)
0	0.00143
1	0.0194
2	0.10489
3	0.2836
4	0.38338
5	0.2073

The probabilities sum to 1.0, so the distribution is correct.

Example
A manufacturer of radios solders a special transformer into each radio. After 5 years, the chance that the transformer is still operational for a given radio is 74%. The company sells 50 radios to the military. What is the chance that precisely 38, 39, or 40 of the radios still work after 5 years?

Solution:

$X \sim \text{Binomial}(n = 50, p = 0.74)$

We must find $\Pr(X = 38 \text{ or } X = 39 \text{ or } X = 40) = f(38) + f(39) + f(40)$.

$$f(38) = \binom{50}{38} \cdot 0.74^{38} \cdot 0.26^{12} = 0.12437$$

$$f(39) = \binom{50}{39} \cdot 0.74^{39} \cdot 0.26^{11} = 0.10891$$

$$f(40) = \binom{50}{40} \cdot 0.74^{40} \cdot 0.26^{10} = 0.08525$$

$\Pr(X = 38 \text{ or } X = 39 \text{ or } X = 40) = 0.31853$

Section 3-8: Continuous Uniform Distribution

The uniform distribution defined over a continuous interval is perhaps the simplest continuous distribution. This distribution assumes that all x-values over a closed interval [a, b] are equally likely. Since the random variable is continuous, a pdf defines the relative likelihood of each x-value. To illustrate the fact that all $x \in [a,b]$ are equally probable, the pdf must be a horizontal segment raised above the x-axis for $x \in [a,b]$. The pdf essentially resembles a square wave or a top hat. A further requirement is that the area under the pdf is 1. Solving for the height of the pdf produces the equation below:

Continuous Uniform Probability Density Function
$$f(x) = \begin{cases} \dfrac{1}{b-a}, & \text{if } a \leq x \leq b \\ 0, & \text{else} \end{cases}$$

For x-values outside the interval [a, b], the pdf will be zero. The interval bounds a and b must be finite, otherwise the pdf would be zero everywhere.

Notation for the Continuous Uniform Distribution
a = lower, real bound
b = upper, real bound

The df for a continuous uniform distribution is linear for $x \in [a,b]$. The cumulative probability will naturally be zero for x-values below 'a' because the pdf is zero up to X = a. The df then climbs in a consistent, linear manner from X = a to X = b. At X = b, the df achieves 1.0 and remains at that maximum value thereafter.

Continuous Uniform Distribution Function

$$F(x) = \begin{cases} 0, & \text{if } x < a \\ \dfrac{x-a}{b-a}, & \text{if } a \leq x \leq b \\ 1, & \text{if } x > b \end{cases}$$

The pdf's and df's for several continuous uniform distributions are plotted below:

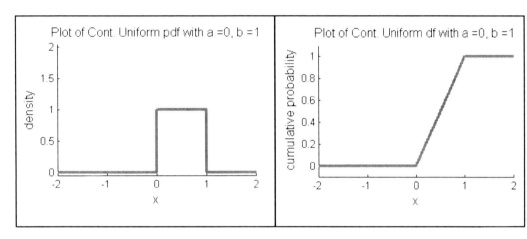

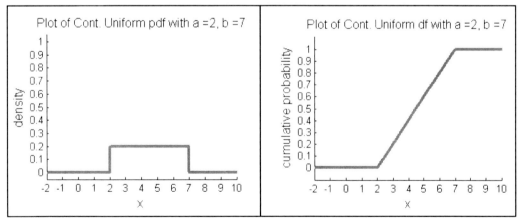

Example
Find the pdf and df for a uniform distribution over the interval (-3, 10).

Solutions:
a = -3, b = 10

$$f(x) = \begin{cases} \dfrac{1}{10-(-3)} = \dfrac{1}{13}, & \text{if } -3 \leq x \leq 10 \\ 0, & \text{else} \end{cases}$$

$$F(x) = \begin{cases} 0, & \text{if } x < -3 \\ \dfrac{x-(-3)}{10-(-3)} = \dfrac{x+3}{13} = \dfrac{1}{13}x + \dfrac{3}{13}, & \text{if } -3 \leq x \leq 10 \\ 1, & \text{if } x > 10 \end{cases}$$

Example
Find the pdf and df for X~Uniform(12, 30).

Solutions:
a = 12, b = 30

$$f(x) = \begin{cases} \dfrac{1}{18}, & \text{if } 12 \leq x \leq 30 \\ 0, & \text{else} \end{cases}$$

$$F(x) = \begin{cases} 0, & \text{if } x < 12 \\ \dfrac{x-12}{18} = \dfrac{1}{18}x - \dfrac{2}{3}, & \text{if } 12 \leq x \leq 30 \\ 1, & \text{if } x > 30 \end{cases}$$

The *standard continuous uniform distribution* arises when a = 0 and b = 1.

Chapter 4: Multivariate Distributions

Multivariate distributions involve two or more variables. The variables could be dependent or independent upon each other. The distribution that covers all variables is called a joint distribution. The joint distribution can be marginalized, revealing the distribution of a proper subset of the available variables. A distribution involving exactly two random variables is called a *bivariate distribution*. Most practical examples will use just two variables. This chapter shows how to find the distribution for one variable which is a function of another random variable with a given distribution. The chapter also covers conditional distributions in which given values of one variable can help predict other variables.

Section 4-1: Basics of Bivariate Distributions

A bivariate distribution involves two variables. Both variables could be discrete, both variables could be continuous, or one variable could be discrete while the other is continuous. Assume that the two variables are X and Y. Every ordered pair (x, y) has a real probability or relative probability of occurrence, depending on whether the variables are discrete or continuous.

Continuous Joint pdf's

If X and Y are both continuous, then the variables have a continuous joint pdf denoted as f(x, y). The joint pdf will resemble a surface with its height at (x, y) representing the relative likelihood that X = x and Y = y in the same experiment. Recall that in the case of a single continuous variable, we could integrate its pdf, and area represented probability. However, in the case of a continuous joint pdf, we must double integrate to find probability, meaning that volume corresponds to probability. Furthermore, to compute the probability that X and Y lie within certain ranges, you must find the volume under the pdf surface and above the x-y plane. A slice or curve through the solid will have zero volume and thus zero probability. Likewise, any given point (x, y) must have zero probability.

The continuous joint pdf must still satisfy all the probability axioms. The function f(x, y) is allowed to be zero or positive, but it cannot be negative. f(x, y) could be arbitrarily large (perhaps around a vertical asymptote). In addition, the double integral of f(x, y) over all possible x- and y-values must be 1.

Rules for a Continuous Joint Probability Density Function

(1.) $f(x, y) \geq 0$

(2.) $\iint_{y\ x} f(x, y)\ dx\ dy = 1$

(3.) Suppose you need to find the chance that (x, y) belongs to set V, where V is a region on the x-y plane. Then,
$$\Pr((x, y) \in V) = \iint_{(x,y)\in V} f(x, y)\ dx\ dy$$

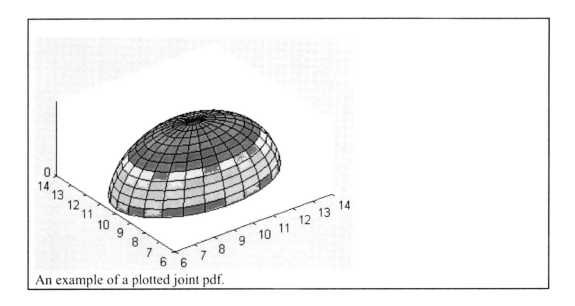

An example of a plotted joint pdf.

Example

X and Y are continuous random variables with this joint pdf:

$$f(x,y) = \begin{cases} ky + 0.2x, & \text{if } 0 < x < 1 \text{ and } 0 < y < 0.75 \\ 0, & \text{else} \end{cases}$$

(a.) Compute k.
(b.) What is the chance that X exceeds 0.5?
(c.) Find $\Pr(X \leq 0.1 \cap Y > 0.25)$

Solutions:
(a.)

$$\int_0^{0.75} \int_0^1 ky + 0.2x \; dx \; dy = \int_0^{0.75} \left[kyx + 0.1x^2 \right]_0^1 dy = \int_0^{0.75} ky + 0.1 \; dy$$

$$= \left[\frac{k}{2} y^2 + 0.1y \right]_0^{0.75} = \frac{k}{2} 0.75^2 + 0.1 * 0.75 = 0.28125k + 0.075$$

The area under the surface must be 1.

$$0.28125k + 0.075 = 1 \quad \rightarrow \quad k = 148/45$$

(b.)

$$\Pr(X > 0.5) = \int_0^{0.75} \int_{0.5}^1 \frac{148}{45} y + 0.2x \; dx \; dy = \int_0^{0.75} \left[\frac{148}{45} yx + 0.1x^2 \right]_{0.5}^1 dy$$

$$= \int_0^{0.75} \left(\frac{148}{45} y + 0.1 \right) - \left(\frac{74}{45} y + 0.025 \right) dy$$

$$= \int_0^{0.75} \frac{74}{45} y + 0.075 \; dy = \left[\frac{37}{45} y^2 + 0.075y \right]_0^{0.75}$$

$$= \frac{37}{45} * 0.75^2 + 0.075 * 0.75 = 0.51875$$

(c.)

$$\Pr(X \leq 0.1 \cap Y > 0.25) = \int_{0.25}^{0.75} \int_0^{0.1} \frac{148}{45} y + 0.2x \; dx \; dy = \int_{0.25}^{0.75} \left[\frac{148}{45} yx + 0.1x^2 \right]_0^{0.1} dy$$

$$= \int_{0.25}^{0.75} \frac{74}{225} y + 0.001 \; dy = \left[\frac{37}{225} y^2 + 0.001y \right]_{0.25}^{0.75} = 0.082722$$

Example

X and Y have this joint pdf:

$$f(x,y) = \begin{cases} c(x+y), & \text{if } 2 < x < 4 \text{ and } -2 < y < 0 \\ 0, & \text{else} \end{cases}$$

(a.) Find the value of c.
(b.) What is the chance that X exceeds 3?

Solutions:

(a.) $c \int_{-2}^{0} \int_{2}^{4} x + y \; dx \; dy = c \int_{-2}^{0} \left[\frac{1}{2}x^2 + yx \right]_{2}^{4} dy$

$= c \int_{-2}^{0} \left(\frac{1}{2}*4^2 + 4y \right) - \left(\frac{1}{2}*2^2 + 2y \right) dy$

$= c \int_{-2}^{0} 8 + 4y - 2 - 2y \; dy = c \int_{-2}^{0} 6 + 2y \; dy = c * \left[6y + y^2 \right]_{-2}^{0}$

$= c*(0 - (-12 + 4)) = c*8$

$8c = 1 \quad \rightarrow \quad c = 1/8$

(b.)

$\Pr(X > 3) = \frac{1}{8} \int_{-2}^{0} \int_{3}^{4} x + y \; dx \; dy = \frac{1}{8} \int_{-2}^{0} \left[\frac{1}{2}x^2 + yx \right]_{3}^{4} dy$

$= \frac{1}{8} \int_{-2}^{0} 8 + 4y - 4.5 - 3y \; dy = \frac{1}{8} \int_{-2}^{0} 3.5 + y \; dy$

$= \frac{1}{8} \left[3.5y + \frac{1}{2}y^2 \right]_{-2}^{0} = \frac{1}{8}(-(-7 + 2)) = 5/8$

Discrete Joint pf's

When X and Y are both discrete, the variables have a discrete joint pf. The discrete joint pf is also denoted f(x, y), and it likewise represents $\Pr(X = x \cap Y = y)$. When f(x, y) is a joint pf, it will not technically define a 3D surface (only continuous functions can form a surface), but the function often forms the outline for a surface. To find the probability of a range of x- and y-values, you must double sum f(x, y) over the valid (x, y) combinations. A discrete joint probability function is usually written in table form. Within the table, rows might correspond to particular x-values, each column is for a certain y-value, and each inner cell gives the value of f(x, y).

A discrete joint probability function will obey the probability axioms. In particular, f(x, y) must always be between 0 and 1, inclusive. Additionally, summing up f(x, y) over all possible pairs (x, y) must produce 1. The quantity of potential (x, y) ordered pairs is allowed to be finite or countably infinite.

Rules for a Discrete Joint Probability Function
(1.) $0 \leq f(x, y) \leq 1$
(2.) $\sum_x \sum_y f(x, y) = 1$
(3.) Suppose you need to find the chance that (x, y) belongs to set V. Then,
$$\Pr((x, y) \in V) = \sum_{(x,y) \in V} \sum f(x, y)$$

Example
The discrete random variables X and Y have the joint pf in the table below:

X \ Y	18	21	24	27	30
15	0.13	0.06	0.03	0.04	0.01
20	0.05	0.04	0.06	0.02	0.03
25	0.07	0.01	0.09	0.07	0.04
30	0.02	0.06	0.04	0.05	0.08

The 3D plot of the joint pf appears here:

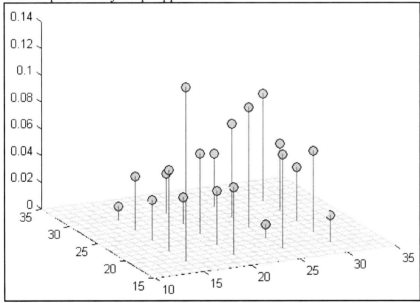

(a.) What is the chance that X = 25 and Y = 27?
(b.) What is the chance that Y = 21?
(c.) Compute $\Pr(X \geq 25 \cap Y \geq 24)$.

Solutions:

(a.) 0.07   (b.) 0.17
(c.) Pr = f(25, 24) + f(25, 27) + f(25, 30) + f(30, 24) + f(30, 27) + f(30, 30) = 0.37

Mixed Joint pf/pdf's

A mixed distribution arises when one variable is discrete and the second variable is continuous. The function f(x, y) is called the mixed pf/pdf. Mixed distributions also obey the three fundamental probability axioms:

Rules for a Mixed Joint pf/pdf
Assume X is discrete and Y is continuous.

(1.) $f(x, y) \geq 0$

(2.) $\int \sum_y \sum_x f(x, y) = 1$

(3.) Suppose you need to find the chance that (x, y) belongs to set V. Then,
$$\Pr((x, y) \in V) = \int_{(x,y) \in V} \sum f(x, y)$$

Example
X is discrete with potential values 5 and 7. Y is continuous with possible values in (0, 4).
The two variables have this joint pf/pdf:

$$f(x, y) = \begin{cases} \dfrac{x\,y}{96}, & \text{if } x \in \{5, 7\} \text{ and } y \in (0, 4) \\ 0, & \text{else} \end{cases}$$

(a.) What is the chance X = 5?
(b.) What is the chance 0 < Y < 3?
(c.) Find the probability that X = 7 and 2 < Y < 4.

Solutions:

(a.) $\Pr(X = 5) = \int_0^4 \dfrac{5}{96} y \, dy = \dfrac{5}{192}\left[y^2\right]_0^4 = \dfrac{80}{192} = \dfrac{5}{12}$

(b.) $\Pr(0 < Y < 3) = \Pr(0 < Y < 3 \cap X = 5) + \Pr(0 < Y < 3 \cap X = 7)$

$= \int_0^3 \dfrac{5}{96} y \, dy + \int_0^3 \dfrac{7}{96} y \, dy = \dfrac{5}{192}\left[y^2\right]_0^3 + \dfrac{7}{192}\left[y^2\right]_0^3$

$$= \frac{45}{192} + \frac{63}{192} = \frac{108}{192} = 9/16$$

(c.) $\Pr(X = 7 \cap 2 < Y < 4) = \int_{2}^{4} \frac{7}{96} y \, dy = \frac{7}{192}\left[y^2\right]_{2}^{4} = 7/16$

The Joint Distribution Function

The joint distribution function is written as F(x, y). This function represents the probability that $X \leq x$ and $Y \leq y$. The joint distribution function exists regardless of whether the variables are discrete or continuous. Evaluating F at the highest possible x- and y-values should produce 1, and evaluating F at the lowest possible x- and y-values should yield 0.

$$\lim_{x \to \infty, y \to \infty} F(x, y) = 1 \qquad \lim_{x \to -\infty, y \to -\infty} F(x, y) = 0$$

To build F(x, y) from the pdf f(x, y), you can double integrate as follows:

$$F(x, y) = \int_{-\infty}^{y} \int_{-\infty}^{x} f(u, v) \, du \, dv$$

Evaluating the second partial derivative of the bivariate distribution function with respect to each variable yields the joint pdf (assuming X and Y are continuous):

$$\frac{\partial^2}{\partial x \partial y}[F(x, y)] = f(x, y)$$

Example
A machine randomly picks an ordered pair (x, y) from a disc of radius 5 centered at the origin of a Cartesian coordinate system. All possible points in the disc have the same chance of being chosen.
(a.) Write the joint pdf between X and Y.
(b.) What is the chance that $1 < X < 3$ and $2 < Y < 4$?

Solutions:
(a.)
The volume of the three-dimensional cylinder must be 1.
The area of the disc is $A = 25\pi$.
The formula for the volume of a cylinder is $V = \pi r^2 h$.

Therefore, $1 = 25\pi h \quad \to \quad h = \dfrac{1}{25\pi}$.

The surface f(x, y) must be level at $\dfrac{1}{25\pi}$ for all points (x, y) in the disc.

$$f(x, y) = \begin{cases} \dfrac{1}{25\pi}, & \text{if } x^2 + y^2 \leq 25 \\ 0, & \text{else} \end{cases}$$

(b.) The area of the square defined by the bounds on X and Y is 4.
The volume of the prism over the square is $4/(25\pi)$.
Therefore, the probability that $1 < X < 3$ and $2 < Y < 4$ is $4/(25\pi)$.

Suppose X and Y are continuous, and you wish to find the probability that $x_1 \leq X \leq x_2$ and $y_1 \leq Y \leq y_2$. In other words, you want to find the volume under f(x,y) for X and Y defined over a rectangular region. A shortcut formula is to compute:

$$\Pr(x_1 \leq X \leq x_2, \ y_1 \leq Y \leq y_2) = F(x_2, y_2) - F(x_1, y_2) - F(x_2, y_1) + F(x_1, y_1)$$

We must add $F(x_1, y_1)$ back in because it is subtracted off twice.

Example
X and Y have this joint pdf:

$$f(x, y) = \begin{cases} 14e^{-2x}e^{-7y}, & \text{if } x > 0 \text{ and } y > 0 \\ 0, & \text{else} \end{cases}$$

(a.) What is the joint df?
(b.) What is $\Pr(X < 0.4 \cap Y < 0.1)$?
(c.) What is $\Pr(X < 1 \cap Y < 1)$?
(d.) Find $\Pr(0.05 < X < 0.25 \text{ and } 0.07 < Y < 0.38)$.

Solutions:
(a.)
$$F(x, y) = \int_0^y \int_0^x 14 e^{-2u} e^{-7v} \, du \, dv = \int_0^y \left[-7 e^{-2u} e^{-7v} \right]_{u=0}^{u=x} dv$$

$$= \int_0^y -7 e^{-2x} e^{-7v} + 7 e^0 e^{-7v} \, dv = \int_0^y (-7 e^{-2x} + 7) e^{-7v} \, dv$$

$$= (-7e^{-2x} + 7) * \left[\left(-\frac{1}{7}\right)e^{-7v}\right]_0^y = (-7e^{-2x} + 7) * \left(-\frac{1}{7}\right) * \left(e^{-7y} - e^0\right)$$

$$= (e^{-2x} - 1) * (e^{-7y} - 1)$$

$$= e^{-2x-7y} - e^{-2x} - e^{-7y} + 1$$

(b.)
$$\Pr(X < 0.4 \cap Y < 0.1) = F(0.4, 0.1) = e^{-2*0.4-7*0.1} - e^{-2*0.4} - e^{-7*0.1} + 1$$
$$= e^{-1.5} - e^{-0.8} - e^{-0.7} + 1$$
$$= 0.27722$$

(c.)
$$\Pr(X < 1 \cap Y < 1) = F(1, 1) = e^{-9} - e^{-2} - e^{-7} + 1 = 0.86388$$

(d.)
F(0.25, 0.38) = 0.36595
F(0.05, 0.38) = 0.08851
F(0.25, 0.07) = 0.15242
F(0.05, 0.07) = 0.03686

Pr(0.05 < X < 0.25 and 0.07 < Y < 0.38) =
  = F(0.25, 0.38) - F(0.05, 0.38) - F(0.25, 0.07) + F(0.05, 0.07)
  = 0.16188

Section 4-2: Computing Marginal Distributions

A marginal distribution describes a proper subset of the variables from the original, full distribution. A marginal distribution of X considers only x-values without including other variables. A marginal distribution can be represented using any of the traditional functions, including a pdf, pf, or df. A marginal pdf for X describes relative probabilities for only X. The marginal pf/pdf of X is denoted $f_X(x)$. The variable with the marginal distribution is in the subscripts of function names. Each marginal distribution has the same properties as a traditional distribution. For example, the total probability under a marginal pdf must still be 1. A marginal distribution could include more than one variable. For instance, if a full joint distribution applies to X, Y, and Z, then a marginal distribution might incorporate these sets of variables: {X}, {Y}, {Z}, {X, Y}, {X, Z}, and {Y, Z}.

Typically, to find a marginal distribution, you first build the marginal pf or pdf from the full multivariate version. Then, you can find the marginal df (e.g., $F_X(x)$) by summing or integrating the pf or pdf.

Univariate Marginal Distributions

Suppose that X and Y are discrete random variables that share a joint distribution. The joint pf for the variables is f(x, y). The pf for just X is $f_X(x)$, and the pf for just Y is $f_Y(y)$. $f_X(x)$ represents the probability that X = x, irrespective of potential y-values. To compute $f_X(x)$, simply sum the joint pf over all possible y-values while holding x constant. A similar rule holds for creating $f_Y(y)$. Note that the summation always occurs with respect to the opposite variable for which the pf is being calculated.

If X and Y are both discrete, then the marginal pf's are:

$$f_X(x) = \sum_y f(x, y) \qquad f_Y(y) = \sum_x f(x, y)$$

The pf for X and Y usually has a 2D tabular format. Suppose x-values lie along the rows, and y-values are along the columns. The sum of the elements in any column gives $f_Y(y)$ for the corresponding y-value in the column. The sum of the elements in any row gives $f_X(x)$ for a particular x-value.

Example
The random variables X and Y have a bivariate distribution with this pf, where each cell represents f(x, y):

		Y		
		3	6	9
	4	0.09	0.04	0.03
X	8	0.16	0.05	0.08
	12	0.18	0.11	0.26

Build the marginal probability functions for X and Y.

Solutions:
The marginal pf for X is:

x	$f_X(x)$
4	0.16
8	0.29
12	0.55

The marginal pf for Y is:

y	$f_Y(y)$
3	0.43
6	0.20

| 9 | 0.37 |

Example
Using the data in the previous example, compute…
(a.) Pr(X = 4 or X = 12)
(b.) Pr(X = 8 | Y = 9)
(c.) Pr(Y = 3 | X = 12)

Solutions:
You can use a marginal pf just like a normal pf.

(a.) Pr(X = 4 or X = 12) = Pr(X = 4) + Pr(X = 12) = 0.16 + 0.55 = 0.71

(b.) $\Pr(X = 8 \mid Y = 9) = \dfrac{\Pr(X = 8 \cap Y = 9)}{\Pr(Y = 9)} = \dfrac{0.08}{0.37} = 0.21622$

(c.) $\Pr(Y = 3 \mid X = 12) = \dfrac{\Pr(Y = 3 \cap X = 12)}{\Pr(X = 12)} = \dfrac{0.18}{0.55} = 0.32727$

Suppose that X and Y are continuous random variables with a joint pdf, f(x,y), connecting them. Rather than summing x- or y-values, an integral must be employed. To find the marginal pdf of X, integrate f(x, y) over all possible y-values. To find the marginal pdf of Y, integrate f(x, y) over all possible x-values.

If X and Y are both continuous, then the marginal pdf's are:

$$f_X(x) = \int_{-\infty}^{\infty} f(x, y)\, dy \qquad f_Y(y) = \int_{-\infty}^{\infty} f(x, y)\, dx$$

Example
X and Y are both continuous with this joint pdf:

$$f(x, y) = \begin{cases} 1/50, & \text{if } 50 < x < 60 \text{ and } 100 < y < 105 \\ 0, & \text{else} \end{cases}$$

What are the marginal pdf's for X and Y?

Solutions:

$$f_X(x) = \int_{100}^{105} \frac{1}{50} \, dy = \left[\frac{1}{50} y\right]_{100}^{105} = 1/10$$

$$f_X(x) = \begin{cases} 1/10, & \text{if } 50 < x < 60 \\ 0, & \text{else} \end{cases}$$

$$f_Y(y) = \int_{50}^{60} \frac{1}{50} \, dx = \left[\frac{1}{50} x\right]_{50}^{60} = 1/5$$

$$f_Y(y) = \begin{cases} 1/5, & \text{if } 100 < y < 105 \\ 0, & \text{else} \end{cases}$$

Example

X and Y share this joint pdf:

$$f(x,y) = \begin{cases} 4680 x^{11}(1-x) y^4 (1-y), & \text{if } 0 < x < 1 \text{ and } 0 < y < 1 \\ 0, & \text{else} \end{cases}$$

Compute the marginal pdf's for X and Y.

Solution:
$$f(x,y) = 4680(x^{11} - x^{12})(y^4 - y^5)$$

$$f_X(x) = \int_0^1 4680(x^{11} - x^{12})(y^4 - y^5) \, dy = 4680(x^{11} - x^{12}) * \left[\frac{1}{5} y^5 - \frac{1}{6} y^6\right]_0^1$$
$$= 4680(x^{11} - x^{12}) * (1/30) = 156(x^{11} - x^{12}), \text{ when } 0 < x < 1$$

$$f_Y(y) = \int_0^1 4680(x^{11} - x^{12})(y^4 - y^5) \, dx = 4680(y^4 - y^5) * \left[\frac{1}{12} x^{12} - \frac{1}{13} x^{13}\right]_0^1$$
$$= 4680(y^4 - y^5) * (1/156) = 30(y^4 - y^5), \text{ when } 0 < y < 1$$

The third case involves X and Y belonging to a mixed bivariate distribution. Suppose X is discrete and Y is continuous. Then, a marginal pf exists for X, and a marginal pdf exists for Y. To find the marginal pf of X, integrate f(x, y) over all the y-values. To find the marginal pdf of Y, sum f(x, y) over all x-values.

If X is discrete and Y is continuous, then

$$f_X(x) = \int_{-\infty}^{\infty} f(x,y) \, dy \qquad f_Y(y) = \sum_x f(x,y)$$

Besides discovering a marginal pf/pdf, you can also calculate a marginal distribution function. A marginal df applies to a proper subset of the available variables. For example, $F_X(x)$ is the marginal df for X and represents the probability that X is less than or equal to X, irrespective of the values of other variables. To find a marginal df for X given the joint df, take the limit of the joint df as the other variables (excluding X) approach positive infinity.

Marginal Distribution Functions
$$F_X(x) = \lim_{y \to \infty} F(x,y) \qquad \text{and} \qquad F_Y(y) = \lim_{x \to \infty} F(x,y)$$

Independence among Random Variables

It is possible for X and Y to be independent or dependent upon each other. When the variables are independent, a chosen value for X places absolutely no constraints on the potential values for Y (and vice versa). On the other hand, variables that are dependent on each other can restrict the values of the other variable.

Independence of random variables can be confirmed or rejected by examining the marginal distributions of the variables. When X and Y are continuous, f(x, y) must be factorable into two functions h₁(x) and h₂(y) in order for the variables to be independent. The function h₁(x) must be proportional to f_X(x), and h₂(y) must be proportional to f_Y(y). The h functions could be the marginal pdf for each variable.

Independence exists between X and Y iff

$$f(x,y) = f_X(x) * f_Y(y) \qquad \text{AND} \qquad F(x,y) = F_X(x) * F_Y(y)$$

Only when X and Y are independent can you work in reverse from marginal distributions and create the multivariate distribution. Otherwise, for variables which are dependent, it is impossible to find the full multivariate distribution given the marginal distributions.

When X and Y are discrete, you can run a few quick tests to determine whether or not the variables are independent. Suppose that the probability function values are organized in a 2D matrix, with x-values along the rows and y-values along the columns.

Choose any two columns, and divide one column element-wise by the second column. If the resulting quotient vector has the same value in all cells, then X and Y are independent. Similarly, if any row divided by another row produces a constant quotient vector, then X and Y are independent.

When X and Y are continuous, a necessary (but not sufficient) condition for X and Y to be independent is that their joint pdf must be nonzero over only a rectangular region. In addition, the rectangle must have sides parallel with the axes of the coordinate system. For instance, X and Y could be independent if f(x, y) > 0 when $5 < X < 8$ and $2 < Y < 3$. A case where X and Y must be dependent on each other occurs when the pdf is positive over a disc centered at the origin. In the case of a disc, choosing a value of X first will restrict the potential range of y-values, since the ordered pair (x, y) must satisfy the equation $x^2 + y^2 \leq r^2$.

Section 4-3: Distributions with More than Two Random Variables

A multivariate distribution could include three or more random variables. Most of the probability principles from the bivariate case are similar in the general multivariate case. A joint pf/pdf, along with a joint distribution function, exist for general multivariate distributions. The joint pf/pdf's and df's still output a single real value.

General Multivariate Probability Function
When $X_1, X_2, \ldots, X_n$ are discrete random variables, the joint pf is:

$$f(x_1, x_2, \ldots, x_n) = \Pr(X_1 = x_1 \cap X_2 = x_2 \cap \ldots \cap X_n = x_n)$$

General Multivariate Distribution Function

$$F(x_1, x_2, \ldots, x_n) = \Pr(X_1 \leq x_1 \cap X_2 \leq x_2 \cap \ldots \cap X_n \leq x_n)$$

You can calculate the full joint pf/pdf from the full joint distribution function by taking the nth partial derivative of the joint df with respect to all variables:

$$f(x_1, x_2, \ldots, x_n) = \frac{\partial^n F}{\partial x_1 \partial x_2 \ldots \partial x_n}$$

The rule for two random variables being independent can be expanded to more than two variables. If the full joint pdf can be factored into the product of n functions, where each function uses only one of the n available variables and no variable is repeated in the functions, then the set of variables must be independent. Likewise, if the full joint df can be factored into the product of n marginal functions (without repeating a variable), then the variables will be independent.

Independence of Many Random Variables
The random variables $X_1, X_2, \ldots, X_n$ are independent iff

$$f(x_1, x_2, \ldots, x_n) = f_{X_1}(x_1) * f_{X_2}(x_2) * \ldots * f_{X_n}(x_n)$$

AND

$$F(x_1, x_2, \ldots, x_n) = F_{X_1}(x_1) * F_{X_2}(x_2) * \ldots * F_{X_n}(x_n)$$

Multivariate Marginal Distributions

The term "marginal" means that you are analyzing a proper subset of the available random variables. If the full joint distribution has three or more random variables, then any analysis using just one or two random variables is considered marginal.

Example
Suppose you have a multivariate distribution with 5 random variables denoted X_1, X_2, X_3, X_4, and X_5.

The marginal joint pdf for just X_1, X_3, and X_5 is:

$$f_{1,3,5}(x_1, x_3, x_5) = \int_{-\infty}^{\infty} \int_{-\infty}^{\infty} f(x_1, x_2, x_3, x_4, x_5) \, dx_2 \, dx_4$$

The marginal joint pdf for X_4 and X_5 is:

$$f_{4,5}(x_4, x_5) = \int_{-\infty}^{\infty} \int_{-\infty}^{\infty} \int_{-\infty}^{\infty} f(x_1, x_2, x_3, x_4, x_5) \, dx_1 \, dx_2 \, dx_3$$

Section 4-4: Determining the Distribution for Y Depending on X

This section describes how to find the distribution for a new random variable, $Y = r(X)$, assuming we already know the distribution for the base variable X. Examples of Y include $Y = X^2$, $Y = X + 5$, and $Y = \lceil X \rceil$. When X is discrete, the distribution for Y will also be discrete. However, when X is continuous, the distribution for Y could be discrete or continuous. As an example of the conversion from continuous to discrete, X could be uniformly distributed on [0, 4], and $Y = \lfloor X \rfloor$. In this case, Y could assume only the values in {0, 1, 2, 3, 4}. The pf/pdf of Y is denoted $g(y)$. We need the pf/pdf of X, called $f(x)$, in order to find the distribution for g.

When X is discrete, you can determine the pf for Y by summing probabilities from X's pf. First, try to list all the potential y-values. Next, determine which x-value(s) map to each y-value within the transformation function r. To compute g(y), sum f(x) for all x-values where r(x) = y.

Converting Discrete X to Discrete Y
$$g(y) = \sum_{x:r(x)=y} f(x)$$

Example
X is a discrete random variable with this pf:

x	f(x)
-3	0.02
-2	0.06
-1	0.18
0	0.23
1	0.04
2	0.07
3	0.40

Let $Y = |X|$. Find the pf for Y.

Solution:
First of all, find all the y-values which can be produced in the transformation function from the available x-values with positive probability. Second, sum all the probabilities f(x) from x-values which yield each y-value to develop g(y).

y	g(y)
0	0.23
1	0.22
2	0.13
3	0.42

Example
Let the random variable X have this pf:

x	f(x)
2	0.05
4	0.24
5	0.09
7	0.11
9	0.21
11	0.14
12	0.16

Let the random variable Y be determined by this function, dependent upon X:

$$y = r(x) = \begin{cases} 1, & \text{if } 0 < x \leq 5 \\ 2, & \text{if } 5 < x \leq 10 \\ 3, & \text{if } 10 < x \leq 20 \end{cases}$$

What is the chance Y = 1 or 3?

Solution:

y	g(y)
1	0.38
2	0.32
3	0.30

$\Pr(Y = 1 \cup Y = 3) = 0.68$

In the case that X is continuous and Y is continuous, discovering the pdf for Y revolves around working with the inverse function of r(x). This method requires f(x) to be positive only over a finite interval of x-values. The function r must have no discontinuities over this interval. Additionally, r can only be decreasing or only increasing over the interval. The inverse of r(x) is s(y). That is, $x = s(y) = r^{-1}(y)$. To find function s, simply begin with the equation $y = r(x)$ and solve for x. Next, follow the equation below to develop g(y):

Formula for the pdf of Y depending on X

$$g(y) = f(s(y)) \cdot \left| \frac{d}{dy}[s(y)] \right|$$

The algorithm is summarized in the succeeding box:

Algorithm for finding the pdf of Y which depends on X
Input: The problem should give the function for Y defined in terms of X. It should also give function f, the pdf of X.
Output: g(y), the pdf of Y
(1.) Write y = r(x) = <function for y defined in terms of x>.
(2.) Solve for x in terms of y. The resulting function is: x = s(y) = $r^{-1}(y)$.
(3.) Place s(y) into function f, obtaining f(s(y)).
(4.) Differentiate s(y) and compute the absolute value of the result.
(5.) Multiply the outputs from steps 3 and 4, forming g(y).
(6.) Determine the legal bounds for Y based on the given legal bounds for X.

Example
Let the random variable X have this pdf:
$$f(x) = \begin{cases} 30x^4 - 30x^5, & \text{if } 0 < x < 1 \\ 0, & \text{else} \end{cases}$$

The random variable Y depends on X according to the equation $Y = r(X) = X^3$. Find the pdf for Y.

Solution:
$$y = x^3 \rightarrow \sqrt[3]{y} = x = s(y)$$

$$f(s(y)) = f(\sqrt[3]{y}) = \begin{cases} 30(\sqrt[3]{y})^4 - 30(\sqrt[3]{y})^5, & \text{if } 0 < y < 1 \\ 0, & \text{else} \end{cases}$$

$$= \begin{cases} 30y^{4/3} - 30y^{5/3}, & \text{if } 0 < y < 1 \\ 0, & \text{else} \end{cases}$$

$$\frac{d}{dy}[s(y)] = \frac{d}{dy}[\sqrt[3]{y}] = \frac{1}{3}y^{-2/3}$$

For $0 < y < 1$,
$$g(y) = f(s(y)) \cdot \left|\frac{d}{dy}[s(y)]\right| = (30y^{4/3} - 30y^{5/3}) * \frac{1}{3}y^{-2/3} = 10y^{2/3} - 10y$$

In the end,
$$g(y) = \begin{cases} 10y^{2/3} - 10y, & \text{if } 0 < y < 1 \\ 0, & \text{else} \end{cases}$$

Example
Suppose X has this pdf:
$$f(x) = \begin{cases} 7e^{-7x}, & \text{if } x > 0 \\ 0, & \text{else} \end{cases}$$

Let $Y = X^2 + 6$. What is the pdf for Y?

Solution:
$$y = r(x) = x^2 + 6 \quad \rightarrow \quad y - 6 = x^2 \quad \rightarrow \quad x = \pm\sqrt{y-6} \quad \rightarrow \quad s(y) = x = \sqrt{y-6}$$

If x > 0, then y > 6. Let y > 6.

$$f(s(y)) = f(\sqrt{y-6}) = 7e^{-7\sqrt{y-6}}$$

$$\frac{d}{dy}[s(y)] = \frac{d}{dy}[\sqrt{y-6}] = \frac{1}{2}(y-6)^{-1/2}$$

$$g(y) = f(s(y)) \bullet \left|\frac{d}{dy}[s(y)]\right| = \begin{cases} \frac{7}{2}e^{-7\sqrt{y-6}} \bullet (y-6)^{-1/2}, & \text{if } y > 6 \\ 0, & \text{else} \end{cases}$$

Example
The random variable X is uniformly distributed on [5, 8]. The random variable Y is defined as $Y = 5X + 9$. What is the pdf for Y?

Solution:
$$f(x) = \begin{cases} 1/3, & \text{if } 5 \leq x \leq 8 \\ 0, & \text{else} \end{cases}$$

If x = 5, then y = 34.
If x = 8, then y = 49.

$$y = 5x + 9 \quad \rightarrow \quad y - 9 = 5x \quad \rightarrow \quad x = \frac{1}{5}y - \frac{9}{5} = s(y)$$

$$f(s(y)) = 1/3$$

$$\frac{d}{dy}[s(y)] = \frac{d}{dy}\left[\frac{1}{5}y - \frac{9}{5}\right] = 1/5$$

$$g(y) = (1/3)*(1/5) = (1/15), \text{ if } 34 \leq y \leq 49$$

> In the end, Y is uniformly distributed on [34, 49].

Example
The random variable X has this pdf:
$$f(x) = \begin{cases} 256x^2 e^{-8x}, & \text{if } x > 0 \\ 0, & \text{else} \end{cases}$$

Let $Y = \sqrt{X}$. Find Y's distribution.

Solution:
$$y = \sqrt{x} \quad \rightarrow \quad x = y^2 = s(y)$$

$$f(s(y)) = 256 * (y^2)^2 * e^{\wedge}[-8y^2] = 256 * y^4 * e^{\wedge}[-8y^2]$$

$$\frac{d}{dy}[s(y)] = \frac{d}{dy}[y^2] = 2y$$

$$g(y) = 256 * y^4 * e^{\wedge}[-8y^2] * 2y = 512 * y^5 * e^{\wedge}[-8y^2]$$

When $x > 0$, $y > 0$.

$$g(y) = \begin{cases} 512 * y^5 * e^{\wedge}[-8y^2], & \text{if } y > 0 \\ 0, & \text{else} \end{cases}$$

Section 4-5: Determining the Distribution for Many Y's Depending on Many X's

This section generalizes the material in the previous section to the case of arbitrarily many X's and arbitrarily many Y's. The X's do not need to belong to a random sample; each X_i could have a unique distribution. The X's may or may not be independent from each other. The X's share a multivariate distribution with joint pf/pdf called f. For simplicity and the sake of most practical examples, we will often assume that n = (number of X's) = (number of Y's) = 2. An r_i function exists for each Y_i random variable such that $Y_i = r_i(X_1, X_2, ..., X_n)$. Inverse functions also exist that produce x-values from the various y-values. The inverse functions are labeled $X_i = s_i(Y_1, Y_2, ..., Y_n)$, so that each s_i function corresponds to one X_i variable. Notice that the s-functions are not direct inverses of the r-functions. Instead, you must solve for each x_i using the provided r_i functions. After obtaining all the s-functions, you must compute the Jacobian of the s-system. Finding the Jacobian first requires building a matrix of partial derivatives. A partial derivative exists for each s-function with respect to each y-variable. The matrix is organized so that s-functions remain constant for each row and y-variables remain constant for each column. To ultimately find the Jacobian, J, compute the determinant of this matrix. The joint pdf between all the y-variables is found with this formula:

Formula for the joint pdf of many Y's depending on many X's
$$g(y_1, y_2, ..., y_n) = f(s_1, s_2, ..., s_n) \bullet |J|$$

Algorithm for finding the joint pdf of multiple Y's which depend on multiple X's
Input: Formulas for all the Y's defined in terms of the X's. The formula for y_i is $r_i(x_1, x_2, ..., x_n)$. The problem should also give the joint pdf between all the X's, $f(x_1, x_2, ..., x_n)$.
Output: The joint pdf for all the Y's, denoted as $g(y_1, y_2, ..., y_n)$.

(1.) Write the functions r defined in terms of the x's, such as:

$$y_1 = r_1(x_1, x_2)$$
$$y_2 = r_2(x_1, x_2)$$

(2.) Solve for each X-variable in terms of the Y-variables. Name the resulting functions s_i. For example:

$$x_1 = s_1(y_1, y_2)$$
$$x_2 = s_2(y_1, y_2)$$

(3.) Find the partial derivative of each s_i function with respect to each individual y_j variable. That is, you need to evaluate $\dfrac{\partial s_i}{\partial y_j}$ for all indices i and j. Place the resulting partial derivatives in a matrix. The ith row of the matrix has all the partial derivatives for s_i, and the jth column has all the partial derivatives with respect to y_j.

For example,

$$\begin{matrix} \dfrac{\partial s_1}{\partial y_1} & \dfrac{\partial s_1}{\partial y_2} \\ \\ \dfrac{\partial s_2}{\partial y_1} & \dfrac{\partial s_2}{\partial y_2} \end{matrix}$$

(4.) Compute the Jacobian of the s_i functions by finding the determinant of the matrix from step 3. The Jacobian is denoted by J. Compute the absolute value of J.

(5.) Plug all the s_i functions from step 2 into f, the joint pdf for the x's. The result will be $f(s_1, s_2, ..., s_n)$.

(6.) Multiply the results from steps 4 and 5, forming $g(y_1, y_2, ..., y_n)$.

Once you obtain the joint pdf g, you will usually want to find the marginal pdf for a particular Y-variable. The first step is to graph the regions where X_1 and X_2 are defined (the plot will use x_1 and x_2 as the axes). Then, using the legal bounds on X_1 and X_2, along with the r-functions, determine the legal bounds for Y_1 and Y_2 and graph the regions. The graphs aid in integrating the joint pdf g. You will need to integrate g with respect to all legal values of Y_2 when finding the marginal pdf of Y_1. The marginal pdf for Y_1 could vary as differing functions define the bounds of Y_1 and Y_2.

Since computing a Jacobian requires calculating a determinant, you should memorize the formulas for determinants of small matrices:

Determinant of a 2-by-2 matrix

Let $M = \begin{bmatrix} a & b \\ c & d \end{bmatrix}$.

The determinant of M = det(M) = $|M|$ = $ad - bc$

Determinant of a 3-by-3 matrix

Let $A = \begin{bmatrix} a_{11} & a_{12} & a_{13} \\ a_{21} & a_{22} & a_{23} \\ a_{31} & a_{32} & a_{33} \end{bmatrix}$. Then,

$\det(A) = |A| = a_{11}a_{22}a_{33} + a_{12}a_{23}a_{31} + a_{13}a_{21}a_{32} - a_{13}a_{22}a_{31} - a_{11}a_{23}a_{32} - a_{12}a_{21}a_{33}$

When the amount of Y's is less than the amount of X's, it is good practice to create dummy Y-variables so that the amounts of X's and Y's are equal. For example, if we are given that $Y_1 = X_1 + X_2$ but a Y_2 variable is not specified, we can create the function r_2 such that $Y_2 = r_2(X_1, X_2) = X_1$. The equal number of X's and Y's ensures that the matrix with partial derivatives, $\frac{\partial s_i}{\partial y_j}$, is square and has a determinant.

Example
The random variable X_1 is uniformly distributed on [2, 4], and the random variable X_2 is uniformly distributed on [14, 18]. X_1 and X_2 are independent. Find the pdf for Y_1, where $Y_1 = X_1 + X_2$.

Solution:
We must find the joint pdf for X_1 and X_2. First, write down the marginal pdf for X_1 and the marginal pdf for X_2. Since X_1 and X_2 are independent, we can find the joint pdf for X_1 and X_2 by multiplying together their marginal pdf's.

$$f_1(x_1) = \begin{cases} 1/2, & \text{if } 2 \leq x_1 \leq 4 \\ 0, & \text{else} \end{cases} \qquad f_2(x_2) = \begin{cases} 1/4, & \text{if } 14 \leq x_2 \leq 18 \\ 0, & \text{else} \end{cases}$$

$$f(x_1, x_2) = \begin{cases} 1/8, & \text{if } 2 \leq x_1 \leq 4 \text{ and } 14 \leq x_2 \leq 18 \\ 0, & \text{else} \end{cases}$$

Write down the r-functions.

$$y_1 = r_1(x_1, x_2) = x_1 + x_2$$
$$y_2 = r_2(x_1, x_2) = x_1$$

Solve for the x-variables, forming the s-functions.

$$y_1 = y_2 + x_2 \quad \rightarrow \quad x_2 = y_1 - y_2$$

$$x_1 = s_1(y_1, y_2) = y_2$$
$$x_2 = s_2(y_1, y_2) = y_1 - y_2$$

Find the partial derivative of each s-function with respect to each y-variable.

$$\frac{\partial s_1}{\partial y_1} = 0 \qquad \frac{\partial s_1}{\partial y_2} = 1$$

$$\frac{\partial s_2}{\partial y_1} = 1 \qquad \frac{\partial s_2}{\partial y_2} = -1$$

$$J = \det\left(\begin{bmatrix} 0 & 1 \\ 1 & -1 \end{bmatrix}\right) = 0 - 1 = -1$$

$$|J| = 1$$

$$g(y_1, y_2) = f(s_1, s_2) \bullet |J| = f(y_2, y_1 - y_2) * 1 = 1/8$$

Find the bounds over which the joint pdf of Y_1 and Y_2 is positive.

$y_2 = x_1 \in [2, 4]$

$y_1 - y_2 = x_2 \in [14, 18]$
$y_1 - y_2 \geq 14$ and $y_1 - y_2 \leq 18$
$-y_2 \geq 14 - y_1$ and $-y_2 \leq 18 - y_1$
$y_2 \leq -14 + y_1$ and $y_2 \geq -18 + y_1$
$-18 + y_1 \leq y_2 \leq -14 + y_1$

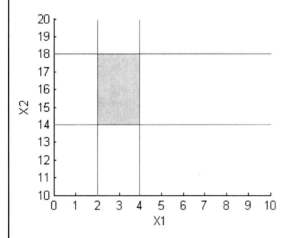

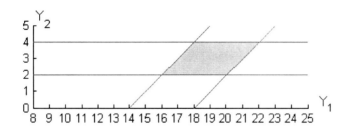

$$g(y_1, y_2) = \begin{cases} 1/8, & \text{if } -18 + y_1 \leq y_2 \leq -14 + y_1 \text{ and } 2 \leq y_2 \leq 4 \\ 0, & \text{else} \end{cases}$$

The marginal pdf for Y_1 will consist of three parts where it is positive.

For $16 \leq Y_1 \leq 18$,

$$f_{Y_1}(y_1) = \int_2^{-14+y_1} (1/8)\, dy_2 = \left[\frac{1}{8}y_2\right]_2^{-14+y_1} = \frac{1}{8}(-14+y_1-2) = -2+\frac{1}{8}y_1$$

For $18 < Y_1 < 20$,

$$f_{Y_1}(y_1) = \int_2^4 (1/8)\, dy_2 = \left[\frac{1}{8}y_2\right]_2^4 = \frac{1}{8}*2 = 1/4$$

For $20 \leq Y_1 \leq 22$,

$$f_{Y_1}(y_1) = \int_{-18+y_1}^4 (1/8)\, dy_2 = \left[\frac{1}{8}y_2\right]_{-18+y_1}^4 = \frac{1}{8}(4+18-y_1) = 2.75 - \frac{1}{8}y_1$$

In the end,

$$f_{Y_1}(y_1) = \begin{cases} -2+\frac{1}{8}y_1, & \text{if } 16 \leq y_1 \leq 18 \\ 1/4, & \text{if } 18 < y_1 < 20 \\ 2.75-\frac{1}{8}y_1, & \text{if } 20 \leq y_1 \leq 22 \\ 0, & \text{else} \end{cases}$$

Example

The random variables X_1 and X_2 share this joint pdf:

$$f(x_1, x_2) = \begin{cases} \frac{1}{5}x_1 + \frac{1}{15}x_2, & \text{if } 0 < x_1 < 3 \text{ and } 0 < x_2 < 1 \\ 0, & \text{else} \end{cases}$$

Another random variable Y_1 is defined as: $Y_1 = X_1 X_2$.
Find the pdf for Y_1.

Solution:

Write down the r-functions.

$y_1 = r_1(x_1, x_2) = x_1 x_2$
$y_2 = r_2(x_1, x_2) = x_1$

Solve for the x-variables, forming the s-functions.

$y_1 = y_2 x_2 \quad \rightarrow \quad x_2 = y_1/y_2$

$$x_1 = s_1(y_1, y_2) = y_2$$
$$x_2 = s_2(y_1, y_2) = y_1/y_2$$

Find the partial derivative of each s-function with respect to each y-variable.

$$\frac{\partial s_1}{\partial y_1} = 0 \qquad \frac{\partial s_1}{\partial y_2} = 1$$

$$\frac{\partial s_2}{\partial y_1} = 1/y_2 \qquad \frac{\partial s_2}{\partial y_2} = -y_1 y_2^{-2}$$

$$J = \det\left(\begin{bmatrix} 0 & 1 \\ 1/y_2 & -y_1 y_2^{-2} \end{bmatrix}\right) = -1/y_2$$

$$|J| = 1/y_2$$

$$g(y_1, y_2) = f(s_1, s_2) \bullet |J| = f(y_2, y_1/y_2) * \frac{1}{y_2} = \left(\frac{1}{5} y_2 + \frac{1}{15} * \frac{y_1}{y_2}\right) * \frac{1}{y_2}$$
$$= \frac{1}{5} + \frac{1}{15} * \frac{y_1}{y_2^2}$$

Find the bounds over which the joint pdf of Y_1 and Y_2 is positive.

$$y_2 = x_1 \in (0, 3)$$

$$y_1/y_2 = x_2 \in (0, 1)$$

$$\frac{y_1}{y_2} > 0 \quad \text{and} \quad \frac{y_1}{y_2} < 1$$
$$y_1 < y_2$$

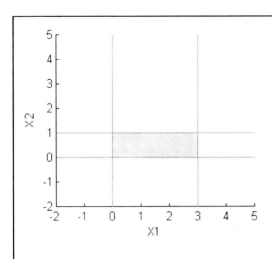

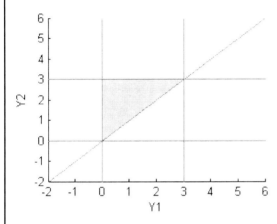

$$g(y_1, y_2) = \begin{cases} \dfrac{1}{5} + \dfrac{1}{15} * \dfrac{y_1}{y_2^2}, & \text{if } 0 < y_1 < 3 \text{ and } y_1 < y_2 < 3 \\ 0, & \text{else} \end{cases}$$

When $0 < y_1 < 3$,

$$f_{Y_1}(y_1) = \int_{y_1}^{3} \frac{1}{5} + \frac{1}{15} y_1 y_2^{-2} \, dy_2$$

$$= \left[\frac{1}{5} y_2 - \frac{1}{15} y_1 y_2^{-1} \right]_{y_2 = y_1}^{y_2 = 3} = \left(\frac{3}{5} - \frac{1}{15} y_1 * 3^{-1} \right) - \left(\frac{1}{5} y_1 - \frac{1}{15} y_1 y_1^{-1} \right)$$

$$= \frac{3}{5} - \frac{1}{45} y_1 - \frac{1}{5} y_1 + \frac{1}{15}$$

$$= \frac{2}{3} - \frac{2}{9} y_1$$

$$f_{Y_1}(y_1) = \begin{cases} \dfrac{2}{3} - \dfrac{2}{9}y_1, & \text{if } 0 < y_1 < 3 \\ 0, & \text{else} \end{cases}$$

Example

X_1 and X_2 are independent random variables. The marginal pdf's for X_1 and X_2 are:

$$f_{X_1}(x_1) = \begin{cases} 5e^{-5x_1}, & \text{if } x_1 > 0 \\ 0, & \text{else} \end{cases} \qquad f_{X_2}(x_2) = \begin{cases} 3e^{-3x_2}, & \text{if } x_2 > 0 \\ 0, & \text{else} \end{cases}$$

Find the pdf for Y_1, where $Y_1 = \dfrac{X_1 + X_2}{2}$.

Solution:

We need to find the joint pdf for X_1 and X_2. Since the two X's are independent, we can multiply their marginal pdf's to get the joint pdf.

$$f(x_1, x_2) = \begin{cases} 15e^{-5x_1 - 3x_2}, & \text{if } x_1 > 0 \text{ and } x_2 > 0 \\ 0, & \text{else} \end{cases}$$

Write down the r-functions.

$$y_1 = r_1(x_1, x_2) = \frac{1}{2}x_1 + \frac{1}{2}x_2$$

$$y_2 = r_2(x_1, x_2) = x_1$$

Solve for the x-variables, forming the s-functions.

$$y_1 = \frac{1}{2}y_2 + \frac{1}{2}x_2 \quad \rightarrow \quad y_1 - \frac{1}{2}y_2 = \frac{1}{2}x_2 \quad \rightarrow \quad 2y_1 - y_2 = x_2$$

$$x_1 = s_1(y_1, y_2) = y_2$$
$$x_2 = s_2(y_1, y_2) = 2y_1 - y_2$$

$$\frac{\partial s_1}{\partial y_1} = 0 \qquad \frac{\partial s_1}{\partial y_2} = 1$$

$$\frac{\partial s_2}{\partial y_1} = 2 \qquad \frac{\partial s_2}{\partial y_2} = -1$$

$$J = \det\left(\begin{bmatrix} 0 & 1 \\ 2 & -1 \end{bmatrix}\right) = 0 - 2 = -2$$

$$|J| = 2$$

$$g(y_1, y_2) = f(s_1, s_2) \bullet |J| = f(y_2, 2y_1 - y_2) * 2 = 30e^{-5y_2 - 3(2y_1 - y_2)}$$
$$= 30e^{-5y_2 - 6y_1 + 3y_2} = 30e^{-2y_2 - 6y_1}$$

Find the bounds over which the joint pdf of Y_1 and Y_2 is positive.

$y_2 = x_1 \in (0, \infty)$
$y_2 > 0$

$2y_1 - y_2 = x_2 \in (0, \infty)$
$2y_1 - y_2 > 0$
$-y_2 > -2y_1$
$y_2 < 2y_1$

We need Y_2 to satisfy $0 < y_2 < 2y_1$

$$g(y_1, y_2) = \begin{cases} 30e^{-2y_2 - 6y_1}, & \text{if } y_1 > 0 \text{ and } 0 < y_2 < 2y_1 \\ 0, & \text{else} \end{cases}$$

When $Y_1 > 0$,

$$f_{Y_1}(y_1) = \int_0^{2y_1} 30e^{-2y_2 - 6y_1} \, dy_2 = 30 * \int_0^{2y_1} e^{-2y_2} * e^{-6y_1} \, dy_2$$

$$= 30e^{-6y_1} * \int_0^{2y_1} e^{-2y_2} \, dy_2 = 30e^{-6y_1} * \left[\left(-\frac{1}{2}\right)e^{-2y_2}\right]_0^{2y_1}$$

$$= 30e^{-6y_1} * \left(-\frac{1}{2}\right) * \left(e^{-2*2y_1} - e^0\right)$$

$$= -15e^{-6y_1} * \left(e^{-2*2y_1} - 1\right)$$

$$= -15e^{-10y_1} + 15e^{-6y_1}$$

$$f_{Y_1}(y_1) = \begin{cases} -15e^{-10y_1} + 15e^{-6y_1}, & \text{if } y_1 > 0 \\ 0, & \text{else} \end{cases}$$

Ordering Variables in a Random Sample

Consider a random sample with n sample points. The values in the random sample can be ordered from smallest to biggest. Let Y_i be the value of the sample point with the ith largest value. That is, Y_1 = minimum sample point and Y_n = maximum sample point. Each Y_i is a random variable with a distribution. Capital F is the distribution function for any sample point. The Y_i's are sample statistics from a random sample.

The distribution function for Y_1 has a straightforward deviation. The probability that Y_1 is at or below y_1 means that at least one x_i is at or below y_1. This probability statement can be reorganized as the probability of the complement that all X_i's are above y_1. The probability that one X_i is above y_1 is $(1 - F(y_1))$. The chance that all X_i's are above y_1 is

$$\Pr(X_1 > y_1 \cap X_2 > y_1 \cap ... \cap X_n > y_1) = \Pr(X_1 > y_1) * \Pr(X_2 > y_1) * ... * \Pr(X_n > y_1)$$
$$= (1 - F(y_1)) * (1 - F(y_1)) * ... * (1 - F(y_1))$$
$$= (1 - F(y_1))^n$$

We can multiply the probabilities of each $X_i > y_1$ because the X_i's are independent. We then need to take the complement of the chance that all X_i's exceed y_1, forming the chance that at least one X_i is at or below y_1. The final formula for Y_1's df is shown here:

Distribution Function for the Minimum in a Random Sample
$$F_1(y_1) = 1 - (1 - F(y_1))^n$$

The distribution function for Y_n also has an intuitive form. The chance that Y_n is at or below y_n means that no X_i can exceed y_n. We need to find the probability that all X_i's are at or below y_n. The chance that one X_i is at or under y_n is just $F(y_n)$. The probability that every X_i is at or under y_n is

$$\Pr(X_1 < y_n \cap X_2 < y_n \cap ... \cap X_n < y_n) = \Pr(X_1 < y_n) * \Pr(X_2 < y_n) * ... * \Pr(X_n < y_n)$$
$$= F(y_n) * F(y_n) * ... * F(y_n)$$
$$= F(y_n)^n$$

The df for Y_n is displayed below:

Distribution Function for the Maximum in a Random Sample
$$F_n(y_n) = F(y_n)^n$$

You can find the pdf of Y_1 or Y_n by differentiating $F_1(y_1)$ or $F_n(y_n)$, respectively. The pdf for Y_1 is found here:

$$f_1(y_1) = \frac{d}{dy_1}\left[1-(1-F(y_1))^n\right] = -n(1-F(y_1))^{n-1} * \frac{d}{dy_1}\left[-F(y_1)\right]$$

$$= n(1-F(y_1))^{n-1} * \frac{d}{dy_1}\left[F(y_1)\right] = n(1-F(y_1))^{n-1} * f(y_1)$$

The pdf for Y_n is developed here:

$$f_n(y_n) = \frac{d}{dy_n}\left[F(y_n)^n\right] = n * F(y_n)^{n-1} * \frac{d}{dy_n}\left[F(y_n)\right]$$

$$= n * F(y_n)^{n-1} * f(y_n)$$

---

pdf for the Minimum in a Random Sample
$f_1(y_1) = n(1-F(y_1))^{n-1} * f(y_1)$

---

pdf for the Maximum in a Random Sample
$f_n(y_n) = n * F(y_n)^{n-1} * f(y_n)$

Example
A random sample has 5 points, each with a continuous uniform distribution over [3, 7].
(a.) What's the chance the smallest sample point is below 3.2?
(b.) Find the probability the greatest sample point is under 5.9.

Solutions:
Each X_i has this pdf and this df:

$$f(x) = \begin{cases} 1/4, & \text{if } 3 \leq x \leq 7 \\ 0, & \text{else} \end{cases}$$

$$F(x) = \begin{cases} 0, & \text{if } x < 3 \\ \frac{1}{4}x - \frac{3}{4}, & \text{if } 3 \leq x \leq 7 \\ 1, & \text{if } x > 7 \end{cases}$$

(a.)
$$\Pr(Y_1 < 3.2) = 1 - (1 - F(3.2))^5 = 1 - \left(1 - \left(\frac{1}{4}*3.2 - \frac{3}{4}\right)\right)^5 = 0.22622$$

(b.)
$$\Pr(Y_n < 5.9) = F(5.9)^5 = \left(\frac{5.9-3}{4}\right)^5 = 0.2003$$

Section 4-6: Conditional Probability Distributions

In an earlier section, we covered the probabilities of conditional events. Recall that the probability of event A given that event B occurred is:

$$\Pr(A \mid B) = \frac{\Pr(A \cap B)}{\Pr(B)}$$

The events can be replaced with random variables. If we are given the value of one random variable, our prediction of other random variables might change. The expression $(X \mid Y)$ is actually a random variable. As in the case of any random variable, $(X \mid Y)$ has a probability distribution. It has a pf/pdf, df, expected value, variance, and so on. The distribution of $(X \mid Y)$ must also satisfy all the probability axioms. For example, the pf/pdf of $(X \mid Y)$ will be positive or zero always. The integral/sum of this pdf will equal 1. The variables X and Y could be both continuous, both discrete, or mixed.

We can convert the conditional probability statement for events into a conditional pf/pdf for random variables. Let's construct the conditional pf/pdf for $(X \mid Y)$. The numerator of the fraction must give the chance that $X = x$ and $Y = y$, which is simply $f(x, y)$. The denominator will provide the probability that $Y = y$ (assuming the value of Y is given), which is really just $f_Y(y)$. The final conditional pf/pdf for $(X \mid Y)$ is:

$$g_1(x \mid y) = \frac{f(x, y)}{f_Y(y)}$$

A similar conditional pf/pdf can be created for Y given the value of X:

$$g_2(y \mid x) = \frac{f(x, y)}{f_X(x)}$$

Example
X and Y are discrete random variables with this joint pf in tabular form:

		Y		
		3	7	9
X	5	0.07	0.04	0.08
	10	0.15	0.06	0.23
	15	0.09	0.02	0.26

(a.) Build the conditional pf of X given that Y = 9.
(b.) Build the conditional pf of Y given that X = 10.

Solutions:
(a.)
$f_Y(9) = 0.08 + 0.23 + 0.26 = 0.57$

$g_1(x = 5 \mid y = 9) = \dfrac{f(5, 9)}{f_Y(9)} = \dfrac{0.08}{0.57} = 0.14035$

$g_1(x = 10 \mid y = 9) = \dfrac{f(10, 9)}{f_Y(9)} = \dfrac{0.23}{0.57} = 0.40351$

$g_1(x = 15 \mid y = 9) = \dfrac{f(15, 9)}{f_Y(9)} = \dfrac{0.26}{0.57} = 0.45614$

x	$g_1(x \mid y = 9)$
5	0.14035
10	0.40351
15	0.45614

(b.)
$f_X(10) = 0.15 + 0.06 + 0.23 = 0.44$

$g_2(y = 3 \mid x = 10) = \dfrac{f(10, 3)}{f_X(10)} = \dfrac{0.15}{0.44} = 0.34091$

$g_2(y = 7 \mid x = 10) = \dfrac{f(10, 7)}{f_X(10)} = \dfrac{0.06}{0.44} = 0.13636$

$g_2(y = 9 \mid x = 10) = \dfrac{f(10, 9)}{f_X(10)} = \dfrac{0.23}{0.44} = 0.52273$

y	$g_2(y \mid x = 10)$
3	0.34091

7	0.13636
9	0.52273

Example
X and Y are continuous random variables with this joint pdf:

$$f(x, y) = \begin{cases} \frac{1}{4}x + 2y, & \text{if } 1 < x < 3 \text{ and } 0 < y < 0.5 \\ 0, & \text{else} \end{cases}$$

(a.) Find the conditional pdf of X given Y.
(b.) Find the conditional pdf of Y given X.
(c.) If we know that Y = 0.3, what is the chance X exceeds 2?
(d.) If we know that X = 1.5, what is the probability Y is below 0.2?

Solutions:
First, let's find the marginal pdf's of X and Y.

$$f_X(x) = \int_0^{0.5} \frac{1}{4}x + 2y \, dy = \left[\frac{1}{4}xy + y^2\right]_{y=0}^{y=0.5} = \frac{1}{8}x + \frac{1}{4}, \text{ when } 1 < x < 3$$

$$f_Y(y) = \int_1^3 \frac{1}{4}x + 2y \, dx = \left[\frac{1}{8}x^2 + 2yx\right]_{x=1}^{x=3} = \frac{9}{8} + 6y - \frac{1}{8} - 2y = 1 + 4y, \text{ when } 0 < y < 0.5$$

(a.)

$$g_1(x \mid y) = \frac{f(x, y)}{f_Y(y)} = \frac{\frac{1}{4}x + 2y}{1 + 4y}, \text{ when } 1 < x < 3 \text{ and } 0 < y < 0.5$$

(b.)

$$g_2(y \mid x) = \frac{f(x, y)}{f_X(x)} = \frac{\frac{1}{4}x + 2y}{\frac{1}{8}x + \frac{1}{4}}, \text{ when } 1 < x < 3 \text{ and } 0 < y < 0.5$$

(c.)

$$g_1(x \mid y = 0.3) = \frac{\frac{1}{4}x + 0.6}{2.2} = \frac{5}{44}x + \frac{3}{11}$$

$$\Pr(X > 2 \mid Y = 0.3) = \int_2^3 \frac{5}{44}x + \frac{3}{11}\, dx = \left[\frac{5}{88}x^2 + \frac{3}{11}x\right]_2^3 = 49/88$$

(d.)

$$g_2(y \mid x = 1.5) = \frac{\frac{3}{8} + 2y}{\frac{7}{16}} = \frac{6}{7} + \frac{32}{7}y$$

$$\Pr(Y < 0.2 \mid X = 1.5) = \int_0^{0.2} \frac{6}{7} + \frac{32}{7}y\, dy = \left[\frac{6}{7}y + \frac{32}{14}y^2\right]_0^{0.2} = 46/175$$

The conditional pf/pdf formulas can be slightly rearranged to show how the joint pf/pdf could be calculated:

Conditional Probability Distribution Corollaries

$$f(x,y) = g_1(x \mid y) * f_Y(y)$$

$$f(x,y) = g_2(y \mid x) * f_X(x)$$

Earlier, we examined Bayes' Theorem as applied to events. Bayes' Theorem also applies to random variables. Recall that under Bayes' Theorem,

$$\Pr(B_j \mid A) = \frac{\Pr(A \mid B_j) * \Pr(B_j)}{\Pr(A)}$$

We need to replace the events with random variables. Let the random variable Y take the place of event A, so that we know the value of Y. Substitute event B_j with the random variable X because X is allowed to take its full spectrum of values and is not fixed. The probability statement becomes:

$$\Pr(X = x_j \mid Y = y) = \frac{\Pr(Y = y \mid X = x_j) * \Pr(X = x_j)}{\Pr(Y = y)}$$

We can replace the probability expressions with pf/pdf functions:

$$g_1(x_j \mid y) = \frac{g_2(y \mid x_j) * f_X(x_j)}{f_Y(y)}$$

A similar formula for Bayes' Theorem exists for $g_2(y|x_j)$.

Independence of Variables in a Conditional Distribution
The random variables X and Y are independent iff

$g_1(x|y) = f_X(x)$ and $g_2(y|x) = f_Y(y)$

Independence Theorems for Expected Values
When X and Y are independent,...

$E(X|Y) = E(X)$
$E(Y|X) = E(Y)$

Chapter 5: Analyzing Distributions

This chapter covers techniques to analyze univariate and multivariate distributions. Two measures of centrality for a random variable include the *median* and *expected value*. The *mode* is another descriptor of the typical value for a random variable. A measure of spread is the *variance*. When we are working with bivariate distributions, we might desire a measure for how well the two variables follow each other. When one variable increases, does the second variable consistently increase, decrease, or appear to follow no pattern? The covariance and correlation coefficient provide measures of how closely two variables are linearly related. Variables that are independent should have a very weak linear relationship. Variables could still be related in a non-linear fashion, however. The chapter concludes with moment-generating functions. Mgf's uniquely describe a distribution. The expected value and variance of any distribution can be quickly calculated if we have the corresponding mgf.

Section 5-1: Expected Value

The expected value of a random variable X is a central, typical value of X. Another term for expected value is "mean." Frequently, but not necessarily, the probability is concentrated around the expected value. If you had to guess the value of X with no prior information, you should choose E(X) as the prediction. The expected value does not have to be a possible value of X. For instance, the expected value of X could be 9.5, while X can really take only integers.

The method for computing the expected value of a random variable depends on whether the variable is discrete or continuous. Calculating the expectation of X essentially requires finding a weighted "average" of all possible x-values, where the weights are the relative probabilities of occurrence for each x-value. In the discrete case, you can compute the expected value of X by summing the product of each x-value and its probability:

Expected Value of a Discrete Random Variable

$$E(X) = \sum_x x \bullet f(x)$$

When X is continuous, summation becomes integration:

Expected Value of a Continuous Random Variable

$$E(X) = \int_{-\infty}^{\infty} x \bullet f(x)\, dx$$

The sum or integral must cover all x-values where f(x) is positive. In the case of a continuous random variable, the bounds of integration might not be $\pm\infty$; instead, the bounds span x-values where f(x) > 0. Additionally, when f(x) > 0 over several disparate intervals, an integral will be needed for each interval.

Example

The random variable X has a binomial distribution with n = 5 and p = 0.15. Find the expected value of X from first principles.

Solution:
First, build the pf table for X:

x	f(x)
0	0.44371
1	0.39150
2	0.13818
3	0.02438
4	0.00215
5	0.00008

Second, multiply each x-value by its probability and sum the results:
$E(X) = 0.3915 + 2*0.13818 + 3*0.02438 + 4*0.00215 + 5*0.00008 = 0.75$

Example
What is the expected value of a Bernoulli distributed random variable X with p = 0.68?

Solution:
Pr(X = 0) = 0.32, Pr(X = 1) = 0.68

$E(X) = 0*0.32 + 1*0.68 = 0.68$

Example
Find the expected value of a uniformly distributed random variable over the interval [30, 70].

Solution:
$$E(X) = \int_{30}^{70} x \bullet \frac{1}{40}\, dx = \left[\frac{1}{80}x^2\right]_{30}^{70} = \frac{1}{80}(70^2 - 30^2) = \frac{4000}{80} = 50$$

When the pdf of X is an even function, E(X) = 0. Exactly half the probability lies below X = 0 and half the probability resides above X = 0. Additionally, the relative probability of every x-value is met with the same relative probability of the negative version of x; f(x) = f(-x).

Example

Shown below is a plot of the pdf $f(x) = \begin{cases} -0.02x^2 + \dfrac{17}{75}, & \text{if } -3 \le x \le 3 \\ 0, & \text{else} \end{cases}$

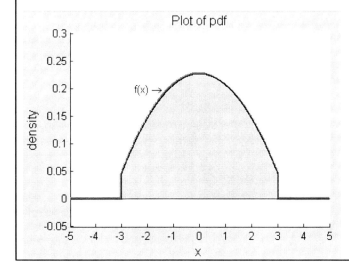

The pdf is an even function; it is symmetric about the vertical axis. The pdf evaluated at x matches the pdf evaluated at –x. To further prove that the expected value of X is 0…

$$E(X) = \int_{-3}^{3} x\left(-0.02x^2 + \frac{17}{75}\right) dx = \int_{-3}^{3} -0.02x^3 + \frac{17}{75}x \, dx$$

$$= \left[\frac{-0.02}{4}x^4 + \frac{17}{150}x^2\right]_{-3}^{3}$$

$$= \left(\frac{-0.02}{4}*81 + \frac{17}{150}*9\right) - \left(\frac{-0.02}{4}*81 + \frac{17}{150}*9\right) = 0$$

Not every distribution has an expected value. For instance, the Cauchy distribution does not have a finite expected value. The tails of the Cauchy pdf approach the x-axis so slowly that x-values near $\pm\infty$ still contribute a sizeable amount of probability. The pdf for the Cauchy distribution is:

$$f(x) = \frac{1}{\pi(1+x^2)}, \text{ for all real x}$$

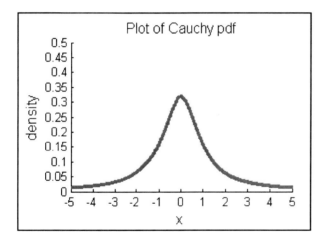

Plot of Cauchy pdf

A simple formula exists to compute the expected value of $aX + b$, where a and b are real-valued constants. The expected value operator is "distributed" throughout the linear expression involving X. The formula for $E(aX + b)$ is developed below:

$$E(aX + b) = E(aX) + E(b) = a*E(X) + E(b) = a*E(X) + b$$

Notice that the expected value of a constant, such as E(b), is just the constant. In the expression $E(aX)$, the coefficient 'a' is pulled outside the expectation and then multiplied by E(X).

Expected Value of a Linear Function of X

$E(aX + b) = a * E(X) + b$

Example

The random variable X has this pdf:

$$f(x) = \begin{cases} 3.75e^{-3.75x}, & \text{if } x > 0 \\ 0, & \text{else} \end{cases}$$

What is $E(6X + 10)$?

Solution:

$$E(X) = \int_0^\infty 3.75 e^{-3.75x} * x \, dx$$

let u = x $\rightarrow$ du = 1 dx

let dv = $e^{-3.75x}$ dx $\rightarrow$ v = $-\dfrac{1}{3.75} e^{-3.75x}$

$$E(X) = 3.75 * \left(\left[-\dfrac{x}{3.75} e^{-3.75x} \right]_0^\infty + \int_0^\infty \dfrac{1}{3.75} e^{-3.75x} \, dx \right)$$

$$= 3.75 * \left(0 + \dfrac{1}{3.75} * \left[-\dfrac{1}{3.75} e^{-3.75x} \right]_0^\infty \right)$$

$$= 3.75 * \left(\dfrac{1}{3.75} * \left(-\dfrac{1}{3.75} \right) * (-1) \right)$$

$$= \dfrac{1}{3.75}$$

$E(6X + 10) = 6 * \dfrac{1}{3.75} + 10 = 11.6$

Simple formulas exist for computing the expected value of a function of X. Suppose Y = r(X) defines a new random variable dependent upon X. We already know the pf or pdf of X, called f(x). The naïve method of finding E(Y) is to find the pf or pdf of Y, called g(y), and then sum or integrate $y \bullet g(y)$ over all possible y-values. However, finding g(y) will take an unnecessary amount of work. A simpler method of finding E(Y) is to replace x with the function r(x) in the previous expected value formulas and then compute the integral or sum of $r(x) \cdot f(x)$ over all x-values:

Expected Value of a Function of a Random Variable
Let $Y = r(X)$ be a function of the random variable X.

If X is continuous, then $E(Y) = \int_{-\infty}^{\infty} r(x) \cdot f(x) \, dx$

If X is discrete, then $E(Y) = \sum_{x} r(x) \cdot f(x)$

Example
The random variable X has a binomial distribution with n = 4 and p = 0.71. A new random variable Y is defined as $Y = X^2$. What is the expected value of Y?

Solution:

x	x^2	f(x)
0	0	0.00707
1	1	0.06926
2	4	0.25437
3	9	0.41518
4	16	0.25412

$E(Y) = E(X^2) = 1*0.06926 + 4*0.25437 + 9*0.41518 + 16*0.25412 = 8.88928$

The expected value is sensitive to outliers. For example, if a probability distribution shows that a very large value of X (large relative to other potential x's) has a chance of occurring, the expected value will shift towards that outlier. Outliers can change the expected value to any arbitrary amount. If the designer of the distribution chose a particular x-value for the outlier and assigned a certain probability to it, he could force E(X) to become any value desired.

The expected value of a random variable differs from the "average." You can only compute the *mathematical average* for a collection of sample points all from the same distribution (that is, they belong to a random sample). To compute an average, simply sum up all the values and divide by the quantity of values. The average gives equal weight of (1/n) to each outcome.

Sample Mean

$$\bar{x} = \frac{\sum_{i=1}^{n} x_i}{n}$$

The sample mean can also be denoted by the random variable $\bar{X}_n$.

Sample Mean Statistics

$$E(\bar{X}_n) = E(X_i) = \mu$$

$$Var(\bar{X}_n) = \frac{Var(X_i)}{n} = \frac{\sigma^2}{n}$$

Probabilities Involving the Sample Mean

Chebyshev Inequality

$$\Pr(|X - E(X)| \geq t) \leq \frac{Var(X)}{t^2}$$

The Chebyshev Inequality says that the probability that X varies from its expected value by t or more units is limited above by Var(X) / t^2. The value t is often a quantity of standard deviations. For instance, t could be 4σ.

The Chebyshev Inequality can provide very abstract statements for any probability distribution. As an example, let X be a random variable, and suppose we want to find the chance that X varies from its expected value by 2 or more standard deviations. That is, let's set t to 2σ. The Chebyshev Inequality becomes

$$\Pr(|X - E(X)| \geq 2\sigma) \leq \frac{\sigma^2}{(2\sigma)^2}$$

$$\downarrow$$

$$\Pr(|X - E(X)| \geq 2\sigma) \leq \frac{\sigma^2}{4\sigma^2}$$

$$\downarrow$$

$$\Pr(|X - E(X)| \geq 2\sigma) \leq \frac{1}{4}$$

The Chebyshev Inequality illustrates the fact that the chance that X varies from E(X) by 2 or more standard deviations is at most 25%. No variable can violate this rule.

Example
A researcher is collecting water from a lake and counting the number of parasites found in each trial. He collects water randomly around the lake so that the water collections form a random sample. The variance in the number of parasites found in a single water analysis is 36. The population mean for the quantity of parasites in one water analysis is still a mystery. However, the researcher hopes to use the sample mean, $\overline{X}_n$, to estimate the population mean, μ. He wants his estimate of the mean to be incorrect by no more than 3 parasites with chance 95%. What is the minimum sample size necessary?

Solution:
$\sigma^2 = 36$, n = ?, μ = ?, t = 3
The random variable for this problem is $\overline{X}_n$.

$\Pr(|\overline{X}_n - \mu| \geq 3) \leq 0.05$

$0.05 = \dfrac{\sigma^2/n}{t^2} \rightarrow 0.05 = \dfrac{36/n}{9} \rightarrow n = 80$

At least 80 water trials must be gathered.

Markov Inequality
$\Pr(X \geq t) \leq \dfrac{E(X)}{t}$

For the Markov Inequality to hold, all the probability for the random variable X must occur for X > 0.

Example
You are given that E(X) = 60. Using the Markov Inequality, what is the chance that X exceeds 90?

Solution:
$\Pr(X \geq 90) \leq \dfrac{60}{90} = \dfrac{2}{3}$

Combining Random Variables under Expectation

To compute the expected value of a sum of random variables, sum the expected value of each random variable. In other words, you can "distribute" the expectation operation to all the inner random variables.

Expected Value of a Sum of Random Variables
Let $X_1, X_2, \ldots, X_k$ be random variables. Regardless of the dependence/independence between the variables,

$$E(X_1 + X_2 + \ldots + X_k) = E(X_1) + E(X_2) + \ldots + E(X_k)$$

The above formula can be generalized to the case where each variable has a coefficient, and an additive term is included:

Expected Value of a Sum of Random Variables with Coefficients and an Additive Term

Let $X_1, X_2, \ldots, X_k$ be random variables, and allow $a_1, a_2, \ldots, a_k$, and b to be real values. Regardless of the dependence/independence between the variables,

$$E(a_1 X_1 + a_2 X_2 + \ldots + a_k X_k + b) = a_1 E(X_1) + a_2 E(X_2) + \ldots + a_k E(X_k) + b$$

Example
Consider three random variables—X, Y, and Z. We know that $E(X) = -8$, $E(Y) = 15$, and $E(Z) = 4$. Calculate $E(2X - Y + 9Z - 12)$.

Solution:
$E(2X - Y + 9Z - 12) = 2*E(X) - E(Y) + 9*E(Z) - 12 = -7$

When the random variables in a collection are all independent, you can compute the expected value of their product by finding the product of their individual expected values. The expected value operator can again be "distributed" among all the variables. The formulas for the expected value of a product of two variables and then arbitrarily many variables are shown here:

Expected Value of XY
Let X and Y be independent random variables. Then,

$$E(XY) = E(X) * E(Y)$$

Expected Value of a Product of Many Random Variables
Let $X_1, X_2, \ldots, X_k$ be independent random variables. Then,

$$E\left(\prod_{i=1}^{k} X_i\right) = \prod_{i=1}^{k} E(X_i)$$

Example
The random variables X, Y, and Z are independent. You are given that $E(X) = 14$, $E(Y) = -5$, and $E(Z) = 6$. Compute...
(a.) E(XYZ)   (b.) E(XY)   (c.) E(YZ)

Solutions:
(a.) $E(XYZ) = E(X) * E(Y) * E(Z) = 14 * (-5) * 6 = -420$
(b.) $E(XY) = E(X) * E(Y) = 14 * (-5) = -70$
(c.) $E(YZ) = E(Y) * E(Z) = -5 * 6 = -30$

Mean Squared Error

The mean squared error (or MSE) provides a measure of how well predicted values of a random variable fit observed values. The "error" is viewed as the expression $(X - d)$, where X is the observed value of the random variable and d is the predicted value. To compute the MSE, you need to find the expected value of the error term squared.

Mean Squared Error Formula
Let X be the random variable of interest and d be the predicted value of X. Then,

$$MSE = E((X - d)^2)$$

To minimize the MSE, you must set d to E(X). If you use E(X) as the prediction, then the variance results for the mean squared error value.

Example
The random variable X is uniformly distributed on the interval [6, 9].
(a.) What is the MSE for using d = 8 as the prediction?

Digital Actuarial Resources   *Comprehensive Probability Review for Actuarial Exams*

(b.) What predicted value will minimize the MSE, and what is the resulting MSE?

Solutions:

(a.) $MSE = E((X-8)^2) = E((X-8)(X-8)) = E(X^2 - 16X + 64)$

Let's find $E(X)$ and $E(X^2)$.

$$E(X) = \int_6^9 \frac{1}{3} x \, dx = \left[\frac{1}{6} x^2\right]_6^9 = 7.5$$

$$E(X^2) = \int_6^9 \frac{1}{3} x^2 \, dx = \left[\frac{1}{9} x^3\right]_6^9 = 57$$

$MSE = E(X^2) - 16E(X) + 64 = 57 - 16*7.5 + 64 = 1$

(b.) The expected value of X will minimize the MSE.

Let $d = E(X) = 7.5$

$MSE = Var(X) = 0.75$

Section 5-2: Conditional Expected Value

When two variables share a bivariate distribution, the given value of one variable can affect the expected value of the other, unknown variable. The conditional expected value of X given $Y = y$ is written as $E(X \mid Y = y)$. The expression $E(X \mid Y)$ is actually a random variable since it depends on Y. The given portion, which is Y, must be a fixed value. Conditional expected values can be computed regardless of whether X and Y are discrete or continuous.

Conditional Expected Value for Continuous Random Variables

When X and Y have a continuous joint distribution, then

$$E(X \mid Y = y) = \int_{-\infty}^{\infty} x \bullet g_1(x \mid y) \, dx$$

and

$$E(Y \mid X = x) = \int_{-\infty}^{\infty} y \bullet g_2(y \mid x) \, dy$$

Conditional Expected Value for Discrete Random Variables

When X and Y have a discrete joint distribution, then

$$E(X \mid Y = y) = \sum_{x} x \bullet g_1(x \mid y)$$

and

$$E(Y \mid X = x) = \sum_{y} y \bullet g_2(y \mid x)$$

The Expected Value of the Expected Value of a Conditional Random Variable

Let X and Y be random variables which may or may not be independent. Then,

$$E(E(X \mid Y)) = E(X)$$

Example
X and Y have this joint probability function:

		Y		
		3	4	5
	2	0.30	0.09	0.06
X	6	0.17	0.02	0.01
	10	0.12	0.16	0.07

Evaluate the following expressions:
(a.) $E(X \mid Y = 3)$ (b.) $E(X \mid Y = 4)$ (c.) $E(Y \mid X = 10)$

Solutions:
(a.) We need to develop the conditional probability function $g_1(x \mid y = 3)$.

$$g_1(x = 2 \mid y = 3) = \frac{f(2, 3)}{f_Y(3)} = \frac{0.30}{0.59} = 30/59$$

$$g_1(x = 6 \mid y = 3) = \frac{f(6, 3)}{f_Y(3)} = \frac{0.17}{0.59} = 17/59$$

$$g_1(x = 10 \mid y = 3) = \frac{f(10, 3)}{f_Y(3)} = \frac{0.12}{0.59} = 12/59$$

Then, compute the conditional expected value using g_1.

$$E(X \mid Y = 3) = \sum_x g_1(x \mid y = 3) * x$$

$$= \left(\frac{30}{59}\right) * 2 + \left(\frac{17}{59}\right) * 6 + \left(\frac{12}{59}\right) * 10 = 282/59 \approx 4.77966$$

(b.)

$$g_1(x = 2 \mid y = 4) = \frac{f(2, 4)}{f_Y(4)} = \frac{0.09}{0.27} = 1/3$$

$$g_1(x = 6 \mid y = 4) = \frac{f(6, 4)}{f_Y(4)} = \frac{0.02}{0.27} = 2/27$$

$$g_1(x = 10 \mid y = 4) = \frac{f(10, 4)}{f_Y(4)} = \frac{0.16}{0.27} = 16/27$$

$$E(X \mid Y = 4) = \sum_x g_1(x \mid y = 4) * x$$

$$= \left(\frac{1}{3}\right) * 2 + \left(\frac{2}{27}\right) * 6 + \left(\frac{16}{27}\right) * 10 = 190/27 \approx 7.03704$$

(c.)

$$g_2(y = 3 \mid x = 10) = \frac{f(10, 3)}{f_X(10)} = \frac{0.12}{0.35} = 12/35$$

$$g_2(y = 4 \mid x = 10) = \frac{f(10, 4)}{f_X(10)} = \frac{0.16}{0.35} = 16/35$$

$$g_2(y = 5 \mid x = 10) = \frac{f(10, 5)}{f_X(10)} = \frac{0.07}{0.35} = 1/5$$

$$E(Y \mid X = 10) = \sum_y g_2(y \mid x = 10) * y$$

$$= \left(\frac{12}{35}\right) * 3 + \left(\frac{16}{35}\right) * 4 + \left(\frac{1}{5}\right) * 5 = 27/7 \approx 3.85714$$

Example
From an earlier example, X and Y are continuous with this joint pdf:

$$f(x,y) = \begin{cases} \dfrac{1}{4}x + 2y, & \text{if } 1 < x < 3 \text{ and } 0 < y < 0.5 \\ 0, & \text{else} \end{cases}$$

We found the conditional pdf of X given Y = y:

$$g_1(x \mid y) = \dfrac{\dfrac{1}{4}x + 2y}{1 + 4y}, \quad \text{when } 1 < x < 3 \text{ and } 0 < y < 0.5$$

Suppose we know that Y = 0.45. What is the expected value of X?

$$E(X \mid Y = 0.45) = \int_1^3 x \bullet \dfrac{\dfrac{1}{4}x + 2*0.45}{1 + 4*0.45}\, dx = \int_1^3 x \bullet \dfrac{\dfrac{1}{4}x + 0.9}{2.8}\, dx$$

$$= \int_1^3 \dfrac{5}{56}x^2 + \dfrac{9}{28}x\, dx = \left[\dfrac{5}{168}x^3 + \dfrac{9}{56}x^2 \right]_1^3 = 173/84 \approx 2.05952$$

The random variable $(X \mid Y)$ also has a variance, which is called the conditional variance. The conditional variance is developed as follows:

$$\begin{aligned}
Var(X \mid Y = y) &= E(((X \mid y) - E(X \mid y))^2) \\
&= E(((X \mid y) - E(X \mid y)) * ((X \mid y) - E(X \mid y))) \\
&= E((X^2 \mid y) - 2*(X \mid y)*E(X \mid y) + E(X \mid y)^2) \\
&= E(X^2 \mid y) - 2*E((X \mid y)*E(X \mid y)) + E(E(X \mid y)^2) \\
&= E(X^2 \mid y) - 2*E(X \mid y)^2 + E(X \mid y)^2 \\
&= E(X^2 \mid y) - E(X \mid y)^2
\end{aligned}$$

MSE for Conditional Distributions

Recall that the mean-squared error in making a prediction is calculated by finding the expected value of the square of the observed value minus the predicted value. That is, in the case of a univariate distribution,

$$MSE = E((X - d)^2)$$

We can extend the MSE formula to the case where we have two variables and already know the value of one variable. The given variable's value could help in predicting the other, unknown variable. Let the given variable be Y and the unknown variable be X. The soon-to-be-observed X term can remain 'X' in the MSE formula because in the end, the observed value of X will not be conditional on Y. The predicted term, however, can use the given y-value. That is, our prediction, d, can really be a function of y, called d(y). Now, the MSE formula becomes:

$$MSE = E((X - d(y))^2)$$

To minimize the conditional MSE, you should use E(X | Y = y) as the prediction. Using this optimal prediction will produce an MSE equal to $Var(X | Y = y)$. In addition, the overall MSE (considering every potential given y-value) equals:

Overall MSE
$E(Var(X | Y))$

Example
The random variables X and Y have the following joint pf:

		Y		
		7	9	12
	5	0.05	0.17	0.19
X	8	0.07	0.22	0.24
	11	0.01	0.02	0.03

(a.) You initially have no given information on Y and need to predict X. Find the predicted value of X that minimizes the MSE of the prediction. What is the resulting MSE?

(b.) You later find that Y = 12. You still need to predict X. What predicted x-value will minimize the MSE, and what is the resulting MSE?

Solutions:
(a.)
Lacking given knowledge of Y, the predicted x-value which will minimize the MSE is E(X).

$f_X(5) = 0.41 \qquad f_X(8) = 0.53 \qquad f_X(11) = 0.06$

prediction = d = E(X) = 0.41*5 + 0.53*8 + 0.06*11 = 6.95

$MSE = Var(X) = E(X^2) - E(X)^2 = 51.43 - 6.95^2 = 3.1275$

(b.)
The predicted x-value that will minimize the MSE is E(X | Y = 12).

$$g_1(x=5 \mid y=12) = \frac{f(5,12)}{f_Y(12)} = 19/46$$

$$g_1(x=8 \mid y=12) = \frac{f(8,12)}{f_Y(12)} = 12/23$$

$$g_1(x=11 \mid y=12) = \frac{f(11,12)}{f_Y(12)} = 3/46$$

$$\text{prediction} = d(Y) = E(X \mid Y=12) = \left(\frac{19}{46}\right)*5 + \left(\frac{12}{23}\right)*8 + \left(\frac{3}{46}\right)*11 = 160/23$$

$$\text{MSE} = Var(X \mid Y=12) = E(X^2 \mid Y=12) - E(X \mid Y=12)^2 = ?$$

$$E(X^2 \mid Y=12) = 25*\left(\frac{19}{46}\right) + 64*\left(\frac{12}{23}\right) + 121*\left(\frac{3}{46}\right) = 1187/23 \approx 51.6087$$

$$\text{MSE} = \left(\frac{1187}{23}\right) - \left(\frac{160}{23}\right)^2 = 1701/529 \approx 3.2155$$

Section 5-3: Median

A *median* of a random variable X is an x-value that divides the probability distribution in half. The value m is a median iff $\Pr(X \leq m) \geq 1/2$ and $\Pr(X \geq m) \geq 1/2$. That is, m is a median when at least half of the probability lies at or below m and at least half of the probability lies at or above m. The 50th percentile is another phrase for median. The median, just as in the case of the expected value, is a centrality measure for a distribution.

Median Definition
The value X = m from a continuous or discrete distribution is a median iff

$\Pr(X \leq m) \geq 1/2$ and $\Pr(X \geq m) \geq 1/2$

All distributions must have at least one median. Some distributions might have infinitely many medians, perhaps over an interval. It is possible for both discrete and continuous distributions to have multiple medians. As long as a value X = m satisfies the mathematical definition of a median, it qualifies as a median.

Example
This problem illustrates a distribution with infinitely many medians.

The random variable X is discrete and can assume the values in the following table with positive probability:

x	f(x)
-4	0.06
0	0.11
4	0.23
8	0.10
12	0.38
16	0.12

Find the median(s) of X.

Solution:
The cumulative distribution function reaches 0.5 precisely at X = 8. Half the probability lies at or below 8, and half the probability lies at or above 8. Therefore, X = 8 must be a median.

Consider other values of X in the interval [8, 12]. The distribution function is constant over the interval [8, 12]. Let's examine the distribution function at X = 9.

$\Pr(X \leq 9) = \Pr(X \leq 8) = 0.5 \geq 1/2$ and $\Pr(X \geq 9) = \Pr(X \geq 12) = 0.5 \geq 1/2$

Therefore, X = 9 is also a median. In fact, for all x-values in [8, 12], the median definition will be satisfied.

Let's also examine the df at X = 12.
$\Pr(X \leq 12) = 0.88 \geq 1/2$ and $\Pr(X \geq 12) = 0.5 \geq 1/2$
As a result, X = 12 is also a median.

In the end, for all $x \in [8, 12]$, x must be a median.

Example
This problem shows a continuous distribution with infinitely many medians.

The pdf below describes the distribution for the random variable X:

$$f(x) = \begin{cases} 0.04x, & \text{if } 0 < x < 5 \\ -0.04x + 0.6, & \text{if } 10 < x < 15 \\ 0, & \text{else} \end{cases}$$

> Find the median(s) of X.
>
> Solution:
>
> $x = 5$ is clearly a median since $\int_0^5 0.04x \, dx = 1/2$.
>
> However, all x-values between 5 and 10, inclusive, are also medians. The pdf is 0 over the x-interval [5, 10], meaning that the cumulative probability is not increasing. Therefore, for any x within [5, 10], the definition of a median is satisfied.
>
> A continuous pdf will typically need to be piecewise in order for it to have an infinite amount of medians.

Unlike the expected value, the median is fairly robust against outliers. Recall that extremely large or small values of X can distort some summary statistics. As long as the outliers have small probabilities of occurrence, they will cause the median to shift only minimally or perhaps remain unchanged.

> **Example**
> Find the median(s) of the distribution for X which has this pdf:
>
> $$f(x) = \begin{cases} 1.8e^{-1.8x}, & \text{if } x > 0 \\ 0, & \text{else} \end{cases}$$
>
> Solution:
> Integrate the pdf from its lowest possible value, 0, up to the mystery median value, m.
>
> $$\int_0^m 1.8e^{-1.8x} \, dx = \left[-e^{-1.8x}\right]_0^m = -e^{-1.8m} + e^0 = 1 - e^{-1.8m}$$
>
> The cumulative probability between 0 and m must be 0.5.
>
> $1 - e^{-1.8m} = 0.5 \quad \to \quad e^{-1.8m} = 0.5 \quad \to \quad \ln(e^{-1.8m}) = \ln(0.5)$
>
> $-1.8m = \ln(0.5) \quad \to \quad m \approx 0.38508$

> **Example**
> Find the median(s) for the distribution of X that has this pf:

$$f(x) = \begin{cases} 0.10, & \text{if } x = 5 \\ 0.35, & \text{if } x = 10 \\ 0.02, & \text{if } x = 15 \\ 0.01, & \text{if } x = 20 \\ 0.08, & \text{if } x = 25 \\ 0.16, & \text{if } x = 30 \\ 0.26, & \text{if } x = 35 \\ 0.02, & \text{if } x = 40 \\ 0, & \text{else} \end{cases}$$

Solution:
The distribution has just one median at x = 25.
To validate that x = 25 is a median,
$\Pr(X \le 25) = 0.56 \ge 1/2$ and $\Pr(X \ge 25) = 0.52 \ge 1/2$

Mean Absolute Error

The mean absolute error (or MAE) provides a technique for determining how close predicted values lie to observed values. The MAE is a counterpart to the MSE encountered earlier. Calculating the MAE involves finding the expected value of the absolute difference between the observed and predicted x-values. To compute the expected value, you can use the expected value formulas for a function of X (e.g., $Y = r(X)$) from earlier.

Mean Absolute Error Formula
Let X be the random variable of interest and d be the predicted value of X. Then,

$MAE = E(|X - d|)$

To minimize the MAE, set d to the median of X.

Example
A random variable has the discrete distribution

x	f(x)
2	0.42
4	0.23
6	0.15
8	0.20

A statistician decides to estimate X with the value 3.

(a.) What is the MAE of the prediction?
(b.) What is the MSE?

Solutions:
(a.)

x	|x – d|	f(x)
2	1	0.42
4	1	0.23
6	3	0.15
8	5	0.20

$MAE = E(|X - d|) = 1*0.42 + 1*0.23 + 3*0.15 + 5*0.2 = 2.1$

(b.)

x	$(x - d)^2$	f(x)
2	1	0.42
4	1	0.23
6	9	0.15
8	25	0.20

$MSE = E((X - d)^2) = 1*0.42 + 1*0.23 + 9*0.15 + 25*0.2 = 7$

Example

X has the pdf $f(x) = \begin{cases} 0.7e^{-0.7x}, & \text{if } x > 0 \\ 0, & \text{else} \end{cases}$

(a.) What predicted value of X minimizes the MAE?
(b.) What predicted value of X minimizes the MSE?

Solutions:
(a.) The median will minimize the MAE.

$\int_0^m 0.7e^{-0.7x}\, dx = 0.5 \rightarrow \left[-e^{-0.7x}\right]_0^m = 0.5 \rightarrow -e^{-0.7m} + e^0 = 0.5$

median = m = 0.99021

(b.) The expected value minimizes the MSE.

$$E(X) = \int_0^\infty x \, 0.7 e^{-0.7x} \, dx$$

let $u = 0.7x \;\to\; du = 0.7\,dx$

let $dv = e^{-0.7x}\,dx \;\to\; v = \dfrac{1}{-0.7}e^{-0.7x}$

$$E(X) = \left[0.7x * \frac{1}{-0.7}e^{-0.7x}\right]_0^\infty + \int_0^\infty e^{-0.7x}\,dx = 0 + \left[\frac{1}{-0.7}e^{-0.7x}\right]_0^\infty$$

$$= \frac{1}{-0.7}(e^{-\infty} - e^0) = \frac{1}{-0.7}(0-1) = \frac{1}{0.7} = 10/7$$

Example
A discrete random variable has this pf:

x	f(x)
3	0.04
6	0.09
9	0.57
12	0.18
15	0.12

What predicted value of X minimizes the MAE, and what is the resulting MAE?

Solutions:
The median will minimize the MAE.
The only median is $X = 9$. So, let's use $d = 9$ for the prediction.

$$MAE = E(|X - d|)$$
$$= 0.04*|3-9| + 0.09*|6-9| + 0.57*|9-9| + 0.18*|12-9| + 0.12*|15-9|$$
$$= 0.04*6 + 0.09*3 + 0.57*0 + 0.18*3 + 0.12*6$$
$$= 1.77$$

Section 5-4: Mode

The *mode* of a random variable is the most common value for the variable. The mode of a continuous random variable X is the x-value at which the pdf attains a global maximum. When X is discrete, the mode is the x-value with the greatest probability. A distribution could be "unimodal" in which it has just one sharp peak at the single mode. A "multimodal" distribution has two or more modes. In a multimodal distribution, each mode occurs at a hill in the pf/pdf, and all the hills attain the same global maximum value.

Example
Find the mode of the continuous random variable X which has this pdf:

$$f(x) = \begin{cases} 364.5x^2 e^{-9x}, & \text{if } x > 0 \\ 0, & \text{else} \end{cases}$$

Solution:
When graphed, this pdf looks like a skewed bell curve. It has a global maximum at the top of its bump. To find the x-value at which the global maximum occurs, differentiate f(x), set the derivative to 0, and then solve for x. The pdf's derivative should be zero at the global maximum.

$$\frac{df}{dx} = 364.5\left(x^2 * (-9)e^{-9x} + 2x * e^{-9x}\right) = 0$$

$$x^2 * (-9)e^{-9x} = -2x * e^{-9x}$$
$$x * (-9) = -2$$

mode = 2 / 9

Example
What is the mode for the random variable X with this pdf?:

$$f(x) = \begin{cases} 168x^5 - 336x^6 + 168x^7, & \text{if } 0 < x < 1 \\ 0, & \text{else} \end{cases}$$

Solution:
Differentiate f(x) with respect to x.

$$\frac{df}{dx} = 840x^4 - 2016x^5 + 1176x^6$$

Set the derivative to 0 and solve for x.

$$840x^4 - 2016x^5 + 1176x^6 = 0$$
$$x^4(840 - 2016x + 1176x^2) = 0$$
$$840 - 2016x + 1176x^2 = 0$$

$$x = \frac{2016 \pm \sqrt{(-2016)^2 - 4*1176*840}}{2*1176}$$
$$\downarrow$$

$$x = \frac{2016 \pm 336}{2352}$$

mode = x = 0.71429

Section 5-5: Variance

The *variance* of a random variable X measures how far X differs from the expected value on average. Variance is a measure of spread for a distribution. The variance of X is abbreviated Var(X) or σ^2. A low variance value means that probability is concentrated around E(X); x-values far from the mean are rare. The variance of a random variable must be greater than or equal to 0. The only case in which Var(X) = 0 is when Pr(X = c) = 1 (all the probability is focused on a single x-value). The variance is said to be defined when the variance is finite. The variance can be swayed by outliers. The variance is guaranteed to exist when X is bounded above and below by finite values.

Variance Formula
$$Var(X) = E((X - E(X))^2) = E(X^2) - E(X)^2$$

The *standard deviation* equals the square root of the variance. The standard deviation is simply denoted by σ.

Example
The random variable X has this pf:

$$f(x) = \begin{cases} 0.07, & \text{if } x = 2 \\ 0.09, & \text{if } x = 4 \\ 0.13, & \text{if } x = 6 \\ 0.29, & \text{if } x = 8 \\ 0.20, & \text{if } x = 10 \\ 0.22, & \text{if } x = 12 \\ 0, & \text{else} \end{cases}$$

Find the variance of X using the classic definition of variance.
What is the standard deviation of X?

Solution:
First, we must find E(X).
$$E(X) = 0.07*2 + 0.09*4 + 0.13*6 + 0.29*8 + 0.20*10 + 0.22*12 = 8.24$$

> Let $U = (X - E(X))^2$. We should calculate u at every corresponding x-value.
>
> | x | u | f(x) |
> |---|---|---|
> | 2 | 38.938 | 0.07 |
> | 4 | 17.978 | 0.09 |
> | 6 | 5.018 | 0.13 |
> | 8 | 0.058 | 0.29 |
> | 10 | 3.098 | 0.20 |
> | 12 | 14.138 | 0.22 |
>
> Now, $Var(X) = E(U) = 8.7428$
> stddev(X) $= \sigma = \sqrt{8.7428} = 2.9568$

Example
The random variable X is discrete and has this pf:

x	f(x)
10	0.15
20	0.48
30	0.06
34	0.11
36	0.20

What is the variance of X? What is the standard deviation of X?

Solution:
First, compute $E(X)$ and $E(X^2)$.

$E(X) = 10*0.15 + 20*0.48 + 30*0.06 + 34*0.11 + 36*0.20 = 23.84$
$E(X^2) = 10^2*0.15 + 20^2*0.48 + 30^2*0.06 + 34^2*0.11 + 36^2*0.20 = 647.36$

$Var(X) = E(X^2) - E(X)^2 = 647.36 - 23.84^2 = 79.0144$
stddev$(X) = \sqrt{Var(X)} = 8.889$

> ### Example
> The random variable X has this pdf:
> $$f(x) = \begin{cases} 4e^{-4x}, & \text{if } x > 0 \\ 0, & \text{else} \end{cases}$$
>
> Compute Var(X).

Solution:
We must first find E(X) and E(X^2).

$$E(X) = \int_0^\infty 4xe^{-4x}\, dx$$

let u = 4x $\quad\to\quad$ du = 4 dx

let dv = e^{-4x} dx $\quad\to\quad$ $v = -\dfrac{1}{4}e^{-4x}$

$$E(X) = \left[-xe^{-4x}\right]_0^\infty + \int_0^\infty e^{-4x}\, dx = 0 + \left[-\dfrac{1}{4}e^{-4x}\right]_0^\infty = -\dfrac{1}{4}e^{-\infty} + \dfrac{1}{4}e^0 = 1/4$$

$$E(X^2) = \int_0^\infty 4x^2 e^{-4x}\, dx$$

let u = 4x^2 $\quad\to\quad$ du = 8x dx

let dv = e^{-4x} dx $\quad\to\quad$ $v = -\dfrac{1}{4}e^{-4x}$

$$E(X^2) = \left[-x^2 e^{-4x}\right]_0^\infty + \int_0^\infty 2xe^{-4x}\, dx = \int_0^\infty 2xe^{-4x}\, dx$$

let u = 2x $\quad\to\quad$ du = 2 dx

let dv = e^{-4x} dx $\quad\to\quad$ $v = -\dfrac{1}{4}e^{-4x}$

$$E(X^2) = \left[-\dfrac{1}{2}xe^{-4x}\right]_0^\infty + \int_0^\infty \dfrac{1}{2}e^{-4x}\, dx = 0 + \dfrac{1}{2}\left[-\dfrac{1}{4}e^{-4x}\right]_0^\infty = -\dfrac{1}{8}(0-1) = 1/8$$

$$Var(X) = E(X^2) - E(X)^2 = \dfrac{1}{8} - \left(\dfrac{1}{4}\right)^2 = 1/16$$

Variance of the Sum of Independent Random Variables
Let $X_1, X_2, \ldots, X_k$ be independent random variables. Assume that the variance of each random variable is defined and finite. Then, the variance of Y, where
$Y = X_1 + X_2 + \ldots + X_k$, is:

$$Var(Y) = \sum_{i=1}^{k} Var(X_i)$$

Example
Two random variables X_1 and X_2 are independent. The marginal pdf's are:

$$f(x_1) = \begin{cases} (1/6)x_1 - (2/3), & \text{if } 9.5 < x_1 < 10.5 \\ 0, & \text{else} \end{cases}$$

$$f(x_2) = \begin{cases} 0.6x_2, & \text{if } 0 < x_2 < \sqrt{10/3} \\ 0, & \text{else} \end{cases}$$

Find the variance of the new random variable $(X_1 + X_2)$.

Solution:
First find the variance of each individual random variable.

$$E(X_1) = \int_{9.5}^{10.5} \frac{1}{6}x_1^2 - \frac{2}{3}x_1 \, dx_1 = \left[\frac{1}{18}x_1^3 - \frac{2}{6}x_1^2\right]_{9.5}^{10.5} = 10.0138889$$

$$E(X_1^2) = \int_{9.5}^{10.5} \frac{1}{6}x_1^3 - \frac{2}{3}x_1^2 \, dx_1 = \left[\frac{1}{24}x_1^4 - \frac{2}{9}x_1^3\right]_{9.5}^{10.5} = 100.36111$$

$$Var(X_1) = 100.36111 - 10.0138889^2 = 0.08314$$

$$E(X_2) = \int_{0}^{\sqrt{10/3}} 0.6x_2^2 \, dx_2 = \left[\frac{0.6}{3}x_2^3\right]_{0}^{\sqrt{10/3}} = 1.21716$$

$$E(X_2^2) = \int_{0}^{\sqrt{10/3}} 0.6x_2^3 \, dx_2 = \left[\frac{0.6}{4}x_2^4\right]_{0}^{\sqrt{10/3}} = 5/3$$

$$Var(X_2) = (5/3) - 1.21716^2 = 0.18519$$

$$Var(X_1 + X_2) = Var(X_1) + Var(X_2) = 0.26833$$

A few useful formulas exist for computing the variance of a linear combination of one or more random variables. Coefficients next to random variables may be taken outside the variance and squared.

Variance of a Linear Function of X
Let $a, b \in R$. Then,

$$Var(aX + b) = a^2 Var(X)$$

Notice that the variance is only affected by a multiplicative factor. The additive amount, b, does not impact the variance. If you simply added b to every x-value, you would anticipate the variance of X to remain unchanged; a constant additive factor does not change the actual distances between x-values.

Variance of a Linear Function of Many Independent X's

Suppose $X_1, X_2, \ldots, X_n$ are n independent random variables.
Also, let $a_1, a_2, \ldots, a_n \in R$ be a set of coefficients, and let $b \in R$ be an additive constant.
Define the random variable Y as: $Y = a_1 X_1 + a_2 X_2 + \ldots + a_n X_n + b$.
Then, the variance of Y is:

$$Var(Y) = a_1^2 Var(X_1) + a_2^2 Var(X_2) + \ldots + a_n^2 Var(X_n)$$

Example

You are given information on three random variables X, Y, and Z:
$Var(X) = 15$, $Var(Y) = 40$, $Var(Z) = 9$

A new random variable, W, is defined as: $W = -7X + 3Y - Z + 10$.
What is the variance of W?

Solution:
$Var(W) = 49 Var(X) + 9 Var(Y) + Var(Z) = 49*15 + 9*40 + 9 = 1,104$

Variance with a Conditional Distribution

Let X and Y be random variables which may or may not be independent. Suppose the distribution of $(X \mid Y)$ exists. Then,

$$Var(X) = E(Var(X \mid Y)) + Var(E(X \mid Y))$$

Section 5-6: Measures of Linear Relationship—Covariance and Correlation

The covariance of two random variables, X and Y, measures the strength of the linear relationship between X and Y. The covariance formula appears below:

Covariance Formula

Let X and Y be two random variables. The covariance between X and Y is defined as:

$$Cov(X,Y) = E((X - E(X))(Y - E(Y))) = E(XY) - E(X) \bullet E(Y)$$

The covariance of two random variables can be any real number. Cov(X, Y) could be negative, positive, or zero. Covariance only measures the linear connection between two variables. As a result, for instance, two variables could have a strong quadratic or cubic relationship, but their linear relationship will register at a weak level. The term "relationship" describes the tendency for the two variables to move in a consistent manner. Two variables are highly linearly correlated when a unit change in X is observed with a consistent change in Y. Correlation does not describe causation; you cannot determine which variable is the independent variable for predicting the other variable without further information.

If two random variables X and Y are independent, then Cov(X, Y) = 0. The converse is not necessarily true; X and Y could have a covariance of zero but be perfectly related in a nonlinear fashion (and thus be dependent on each other).

Example
X and Y are discrete random variables with this joint probability function:

$f_{X,Y}(x, y)$		Y				Total
		12	13	14	15	
X	4	0.13	0.01	0.08	0.07	0.29
	5	0.06	0.12	0.02	0.03	0.23
	6	0.04	0.01	0.14	0.05	0.24
	7	0.02	0.01	0.05	0.16	0.24
Total		0.25	0.15	0.29	0.31	1.00

Compute the covariance between X and Y.
Are X and Y independent?

Solutions:
$E(XY) = 74.7$
$E(X) = 4*0.29 + 5*0.23 + 6*0.24 + 7*0.24 = 5.43$
$E(Y) = 12*0.25 + 13*0.15 + 14*0.29 + 15*0.31 = 13.66$

$E(X)*E(Y) = 74.1738 \neq E(XY)$
∴ X and Y are dependent.

$Cov(X,Y) = E(XY) - E(X)*E(Y) = 74.7 - 5.43*13.66 = 0.5262$
The fact that $Cov(X,Y) \neq 0$ provides further proof that X and Y are dependent.

Example
An experimental process selects a point (x, y) at random over the rectangle $-2 \leq x \leq 4$ and $1 \leq y \leq 3$. What is the covariance between X and Y?

Digital Actuarial Resources   Comprehensive Probability Review for Actuarial Exams

Solution:
At first sight, X and Y don't appear to have any linear relationship. X is selected completely independently from Y; a chosen value of X has no impact on the valid range of Y (and vice versa). A value of X is simply chosen from [-2, 4], and a value of Y is picked from [1, 3]. Let's confirm that X and Y are independent by showing that $Cov(X,Y) = 0$.

$$E(XY) = \int_1^3 \int_{-2}^4 \frac{1}{6*2} * xy \, dx \, dy = \int_1^3 \left[\frac{1}{24} x^2 y \right]_{x=-2}^{x=4} dy = \int_1^3 \frac{1}{24} y(16-4) \, dy = \int_1^3 \frac{1}{2} y \, dy$$

$$= \left[\frac{1}{4} y^2 \right]_1^3 = \frac{1}{4}(9-1) = 2$$

$$E(X) = \frac{-2+4}{2} = 1 \qquad E(Y) = \frac{1+3}{2} = 2$$

$$Cov(X,Y) = 2 - 1*2 = 0$$

Remember that a necessary condition for X and Y to be independent is that their joint probability is nonzero only over a rectangular region.

Covariance between X and X
The covariance of a random variable X with itself is simply Var(X):

$$Cov(X,X) = E((X - E(X))(X - E(X))) = E((X - E(X))^2) = Var(X)$$

The covariance operator is also commutative, so the order of operands does not matter.

Commutative Property of the Covariance
$Cov(X, Y) = Cov(Y, X)$

The covariance value alone does not reveal how closely X and Y are linearly related. If X and Y individually have large variances, then Cov(X, Y) could be relatively large. Correspondingly, Cov(X, Y) depends on how large X and Y typically are. The relationship between X and Y must be "normalized." You can normalize the relationship by dividing the covariance of X and Y by the product of their individual standard deviations. The resulting correlation value is ρ, which is called the *correlation coefficient*.

Correlation Formula
Let X and Y be random variables. The correlation between X and Y is:

$$\rho = \frac{Cov(X,Y)}{\sigma_X \sigma_Y}$$

Rho has a limited span of potential values. The correlation coefficient must be between -1 and 1, inclusive. A value near zero means that X and Y have a weak linear relationship. When $\rho = +1$, X and Y have a perfect positive linear relationship (the slope of a fitted line will be positive). When $\rho = -1$, the variables have a perfect negative linear relationship (the slope of the fitted line would be negative). Notice that Cov(X, Y) and ρ have the same sign (they're either both positive or both negative). The diagram below gives a few examples of plots between two variables and the general linear relationship between the variables.

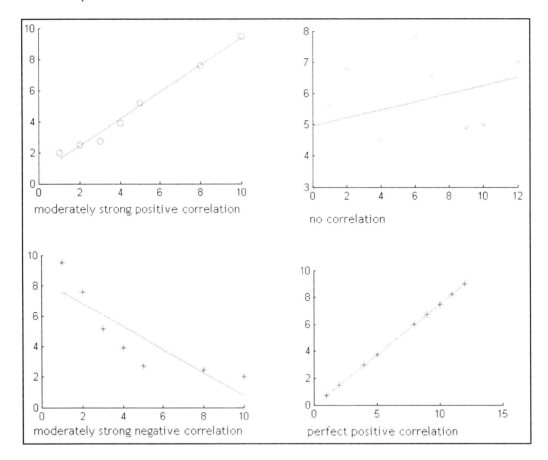

Example
Using the previous joint distribution between discrete variables X and Y, find the correlation between X and Y. How strong is the linear relationship between the variables?

Solution:
We need to find the standard deviations for X and Y.
$E(X^2) = 4^2 * 0.29 + 5^2 * 0.23 + 6^2 * 0.24 + 7^2 * 0.24 = 30.79$
$Var(X) = 30.79 - 5.43^2 = 1.3051$
$\sigma_X = 1.14241$

$E(Y^2) = 12^2 * 0.25 + 13^2 * 0.15 + 14^2 * 0.29 + 15^2 * 0.31 = 187.94$
$Var(Y) = 187.94 - 13.66^2 = 1.3444$
$\sigma_Y = 1.15948$

$$\rho = \frac{Cov(X,Y)}{\sigma_X \sigma_Y} = \frac{0.5262}{1.14241 * 1.15948} = 0.39725$$

The correlation coefficient shows that X and Y have a moderately weak, positive linear relationship.

Example
You are given the following traits about X and Y:
$E(X) = 12$, $E(X^2) = 190$, $E(Y) = 15$, $E(Y^2) = 260$, $E(XY) = 200$.
Find Cov(X, Y) and the correlation coefficient for the relationship. Describe the relationship between X and Y.

Solutions:
$Var(X) = 190 - 144 = 46 \rightarrow \sigma_X = \sqrt{46}$
$Var(Y) = 260 - 225 = 35 \rightarrow \sigma_Y = \sqrt{35}$

$Cov(X,Y) = 200 - 12 * 15 = 20$

$$\rho = \frac{20}{\sqrt{46}\sqrt{35}} = 0.49844$$

X and Y are positively related, but the linear relationship is only moderate in strength.

Covariance with Coefficients
Let X and Y be random variables which may or may not be independent. Also, let $a, b \in R$. Then,

$$Cov(aX, bY) = abCov(X, Y)$$

Variance of a Linear Function of X and Y
Let X and Y be two random variables which may or may not be independent. Then,

$$Var(aX + bY + c) = a^2 Var(X) + b^2 Var(Y) + 2ab Cov(X, Y)$$

Specific Examples of the Variance for a Linear Function of X and Y
$$Var(X + Y) = Var(X) + Var(Y) + 2Cov(X, Y)$$
$$Var(X - Y) = Var(X) + Var(Y) - 2Cov(X, Y)$$

Example
You are given the following information about X and Y:
The correlation coefficient between X and Y is 0.8.
$E(X) = 4$, $E(X^2) = 24$, $E(Y) = 6$, $E(Y^2) = 48$.

Calculate…
(a.) $Var(X - Y)$ (b.) $Var(X + 5Y)$ (c.) $Var(8X + 7Y - 50)$

Solutions:
First, find Var(X), Var(Y), and the individual standard deviations.
Then, solve for Cov(X, Y).

$$Var(X) = 24 - 4^2 = 8 \quad \rightarrow \quad \sigma_X = \sqrt{8}$$
$$Var(Y) = 48 - 6^2 = 12 \quad \rightarrow \quad \sigma_Y = \sqrt{12}$$

$$0.8 = \frac{Cov(X, Y)}{\sqrt{8}\sqrt{12}} \quad \rightarrow \quad Cov(X, Y) \approx 7.83836$$

(a.) $Var(X - Y) = 8 + 12 + 2(1)(-1)Cov(X, Y) = 20 - 2 * 7.83836 \approx 4.32327$

(b.) $Var(X + 5Y) = Var(X) + 25Var(Y) + 2(1)(5)Cov(X, Y)$
$\qquad = 8 + 25 * 12 + 10 * 7.83836 = 386.3836$

(c.) $Var(8X + 7Y - 50) = 64Var(X) + 49Var(Y) + 2(8)(7)Cov(X, Y)$
$\qquad = 64 * 8 + 49 * 12 + 112 * 7.83836 = 1,977.89632$

Variance of the Sum of Many Random Variables
Let X1, X2, ..., Xn be random variables which may or may not be independent. The variance of their sum is:

$$Var(X_1 + X_2 + ... + X_n) = \sum_{i=1}^{n} Var(X_i) + \sum\sum_{i \neq j} Cov(X_i, X_j)$$

Whenever X_i and X_j are independent, you can disregard the covariance between the variables because it will be 0.

Section 5-7: Moment-Generating Functions

The moment-generating function is used, of course, for finding moments. The function provides an elegant way to determine any moment of a random variable. The kth moment of X equals $E(X^k)$, where k is a positive integer. Once you have moments, you can also compute anything that depends on moments, such as a variance or covariance. Every distribution has a corresponding moment-generating function. The Greek letter ψ is reserved for mgf's.

Moment-Generating Function Formula
$$\psi(t) = E(e^{tX})$$

In the mgf formula, t is an extra real-valued variable. We are typically only concerned about the behavior of the mgf around t = 0. You can compute the mgf for X by using the properties for the expectation of a function of X. You can first assign $r(X) = e^{tX}$. Then, multiply r(X) by the pf (or pdf) of X and sum (or integrate) the product over all possible x-values.

Finding moments of X involves differentiating ψ with respect to t and then evaluating the derivative at t = 0. For instance, to find the first moment of X, which is E(X), compute the first derivative of ψ and then plug 0 into the derivative. That is, $E(X) = \psi'(0)$.

k^{th} Moment Formula
$$E(X^k) = \psi^{(k)}(0)$$

Digital Actuarial Resources   *Comprehensive Probability Review for Actuarial Exams*

To compute the kth derivative of ψ, you unfortunately need to find all the previous derivatives and then evaluate $\psi^{(k)}$ at t = 0. The kth moment is defined when $E(|X|^k) < \infty$.

Example
The random variable X has a binomial distribution with parameters n = 4 and p = 0.28.
(a.) Find the mgf for X.
(b.) Compute the first and second moments of X.
(c.) Find the variance of X.

Solution:
(a.)

X	e^{tx}	f(x)
0	1	0.26874
1	e^t	0.41804
2	e^{2t}	0.24386
3	e^{3t}	0.06322
4	e^{4t}	0.00615

$\psi(t) = E(e^{tX})$
$= 0.26874 + 0.41804 * e^t + 0.24386 * e^{2t} + 0.06322 * e^{3t} + 0.00615 * e^{4t}$

(b.)
$\psi'(t) = 0.41804 * e^t + 0.48772 * e^{2t} + 0.18966 * e^{3t} + 0.0246 * e^{4t}$
$E(X) = \psi'(0) = 0.41804 * e^0 + 0.48772 * e^0 + 0.18966 * e^0 + 0.0246 * e^0$
$= 1.12002$

$\psi''(t) = 0.41804 * e^t + 0.97544 * e^{2t} + 0.56898 * e^{3t} + 0.0984 * e^{4t}$
$E(X^2) = \psi''(0) = 0.41804 * e^0 + 0.97544 * e^0 + 0.56898 * e^0 + 0.0984 * e^0$
$= 2.06086$

(c.)
$Var(X) = E(X^2) - E(X)^2 = 2.06086 - 1.12002^2 = 0.80642$

Note that the variance and moments from part b are slightly off from their actual values due to rounding.

Example
Any moment-generating function evaluated at t = 0 must yield 1. Notice that
$\psi(0) = E(e^{0*X}) = E(1) = 1$.

A moment-generating function has a one-to-one correspondence with a particular distribution. If you are given a moment-generating function, there is one and only one distribution that the mgf describes. If two random variables individually have the same mgf, then they must have the same distribution. The template distributions encountered in later chapters each have a common equation for their mgf's. If you can find the mgf for a random variable, you can often match it with the corresponding template distribution.

You may only be able to compute the first few moments of a random variable. If the kth moment of X exists, then you can be guaranteed that the 1st, 2nd, ..., $k-1^{th}$ moments are also defined. However, as soon as you reach a kth moment which is undefined, all the higher moments will also be undefined. In the case that the random variable is bounded, all the moments are guaranteed to be defined. Nonetheless, X is not required to be bounded to have all possible moments. A firm requirement for X to have all possible moments is that its mgf evaluates to a real value for every t-value in an interval that includes t = 0.

Example
A random variable X has this pdf:

$$f(x) = \begin{cases} 32x^2 e^{-4x}, & \text{if } x > 0 \\ 0, & \text{else} \end{cases}$$

(a.) What is the mgf of X?
(b.) Find the first and second moments of X.
(c.) What is the variance of X?

Solutions:
(a.)

$$\psi(t) = E(e^{tX}) = \int_0^\infty e^{tx} * 32x^2 e^{-4x} \, dx = 32 \int_0^\infty e^{x(t-4)} x^2 \, dx$$

let $u = x^2$ $\rightarrow$ $du = 2x \, dx$
let $dv = e^{x(t-4)} \, dx$ $\rightarrow$ $v = \dfrac{1}{t-4} e^{x(t-4)}$

$$\psi(t) = 32 \left(\left[\frac{1}{t-4} e^{x(t-4)} x^2 \right]_{x=0}^{x=\infty} - \frac{2}{t-4} \int_0^\infty x e^{x(t-4)} \, dx \right)$$

$$= 32 \left(\frac{1}{t-4} e^{\infty(t-4)} \infty^2 - \frac{2}{t-4} \int_0^\infty x e^{x(t-4)} \, dx \right)$$

Since any derivative of ψ at t = 0 will still require the calculation of $e^{\infty(t-4)}$, and $e^{-\infty} = 0$, we can set $\frac{1}{t-4} e^{\infty(t-4)} \infty^2$ to 0.

$$\psi(t) = -\frac{64}{t-4} \int_0^\infty x e^{x(t-4)} \, dx$$

let u = x → du = 1 dx
let dv = $e^{x(t-4)}$ dx → v = $\frac{1}{t-4} e^{x(t-4)}$

$$\int_0^\infty x e^{x(t-4)} \, dx = \left[\frac{1}{t-4} e^{x(t-4)} x\right]_{x=0}^{x=\infty} - \int_0^\infty \frac{1}{t-4} e^{x(t-4)} \, dx$$

$$= \frac{1}{t-4} e^{\infty(t-4)} \infty - \left[\frac{1}{(t-4)^2} e^{x(t-4)}\right]_{x=0}^{x=\infty}$$

$$= \frac{1}{t-4} e^{\infty(t-4)} \infty - \left(\left(\frac{1}{(t-4)^2} e^{\infty(t-4)}\right) - \left(\frac{1}{(t-4)^2} e^{0(t-4)}\right)\right)$$

$$= \frac{1}{t-4} e^{\infty(t-4)} \infty - \frac{1}{(t-4)^2} e^{\infty(t-4)} + \frac{1}{(t-4)^2}$$

$$= \frac{1}{(t-4)^2}$$

$$\psi(t) = -\frac{64}{t-4} * \frac{1}{(t-4)^2} = \frac{-64}{(t-4)^3} = \frac{-4^3}{(t-4)^3} = \left(\frac{-4}{t-4}\right)^3 = \left(\frac{4}{4-t}\right)^3 = 64(4-t)^{-3}$$

(b.)
first moment = $E(X) = \psi'(t=0)$

$$\psi'(t) = \frac{d}{dt}\left[64(4-t)^{-3}\right] = 64(-3)(4-t)^{-4}(-1) = 192(4-t)^{-4}$$

$$E(X) = 192(4-0)^{-4} = \frac{192}{256} = 0.75$$

second moment = $E(X^2) = \psi''(t=0)$

$$\psi''(t) = 192(-4)(4-t)^{-5}(-1) = 768(4-t)^{-5}$$

Digital Actuarial Resources — Comprehensive Probability Review for Actuarial Exams

$$E(X^2) = 768(4-0)^{-5} = \frac{768}{1024} = 0.75$$

(c.)
$$Var(X) = 0.75 - 0.75^2 = 0.1875 = 3/16$$

Variance Written in Terms of Moments
$Var(X) = \psi''(0) - (\psi'(0))^2$

Another brand of moment is the kth central moment. The second central moment is the variance.

k^{th} Central Moment Formula
Let k be a positive integer. The kth central moment of X is:
$E((X - E(X))^k)$

Example
What is the first central moment for any random variable X?

Solution:
first central moment = $E(X - E(X)) = E(X) - E(E(X)) = E(X) - E(X) = 0$

Moment-Generating Function for a Linear Combination of X
Suppose ψ_X defines the mgf for the random variable X.
Suppose another random variable is defined as Y = aX + b.
The mgf for Y is:

$$\psi_Y(t) = e^{bt} * \psi_X(at)$$

Moment-Generating Function for a Sum of Independent Random Variables
Suppose that $X_1, X_2, \ldots, X_n$ are independent random variables, each of which has a moment-generating function $\psi_i(t)$. Then, the moment-generating function for $Y = (X_1 + X_2 + \ldots + X_n)$ is:

$$\psi_Y(t) = \prod_{i=1}^{n} \psi_i(t)$$

Example
X has a uniform distribution with this pdf:
$$f_X(x) = \begin{cases} \dfrac{1}{5}, & \text{if } 25 < x < 30 \\ 0, & \text{else} \end{cases}$$

Find the mgf of X.

Solution:
$$\psi_X(t) = E(e^{tX}) = \int_{25}^{30} \frac{1}{5} e^{tx}\, dx = \left[\frac{1}{5t} e^{tx}\right]_{x=25}^{x=30} = \frac{1}{5t}\left(e^{30t} - e^{25t}\right)$$

Unfortunately, this mgf is somewhat useless. First of all, the mgf is undefined at t = 0. If you differentiate the mgf any number of times, trying to evaluate the resulting derivative at t = 0 will produce more errors.

Example
X and Y are continuous random variables with the marginal pdf's displayed below. X and Y are independent.

$$f_X(x) = \begin{cases} e^{-x}, & x > 0 \\ 0, & \text{else} \end{cases} \qquad f_Y(y) = \begin{cases} 10e^{-10y}, & y > 0 \\ 0, & \text{else} \end{cases}$$

Let Z = X + Y.

(a.) What is the mgf for Z?
(b.) Find E(Z).

Solution:
(a.)
$$\psi_X(t) = E(e^{tX}) = \frac{1}{1-t}$$

$$\psi_Y(t) = E(e^{tY}) = \frac{10}{10-t}$$

Since X and Y are independent, we can find the mgf for Z by multiplying the mgf's for X and Y.

$$\psi_Z(t) = \psi_X(t) * \psi_Y(t) = \frac{1}{1-t} * \frac{10}{10-t} = 10(1-t)^{-1}(10-t)^{-1}$$

(b.)
$$\psi_Z'(t) = 10\left((1-t)^{-1} * (-1)(10-t)^{-2}(-1) + (-1)(1-t)^{-2}(-1) * (10-t)^{-1}\right)$$

$$\psi_Z'(0) = 10\left((1-0)^{-1} * (10-0)^{-2} + (1-0)^{-2} * (10-0)^{-1}\right)$$
$$= 10\left(1*10^{-2} + 1*10^{-1}\right) = 1.1 = E(Z)$$

Chapter 6: Common Discrete Distributions

A variety of discrete distributions with formal names exist to describe common phenomena in the field of probability. Each discrete distribution has an expected value, variance, and unique mgf. Every template distribution also has a small set of parameters which you can adjust to fit your experiment. The discrete distributions covered in this chapter include the Bernoulli, binomial, Poisson, negative binomial, geometric, hypergeometric, and multinomial distributions.

Section 6-1: Bernoulli and Binomial Distributions Revisited

Bernoulli Distribution

As discussed earlier, a Bernoulli distributed random variable can take only the values 0 (failure) or 1 (success). The parameter p represents the chance that the variable equals 1.

Example
Derive the expected value and variance of the Bernoulli distribution using first principles.

Solution:
Build a simple table of x-values and probabilities.

x	x^2	f(x)
0	0	1 – p
1	1	p

$E(X) = 0 * f(0) + 1 * f(1) = f(1) = p$

$E(X^2) = 0 * f(0) + 1 * f(1) = p$

$Var(X) = E(X^2) - E(X)^2 = p - p^2 = p(1-p) = pq$

Example
Derive the moment-generating function for the Bernoulli distribution.

Solution:
$\psi(t) = E(e^{tX}) = e^{t*0} * f(0) + e^{t*1} * f(1) = e^0 * (1-p) + e^t * p = q + e^t * p$

Formulas for the Bernoulli Distribution

$E(X) = p$

$Var(X) = pq$

$\psi(t) = pe^t + q$

The kth moment of a Bernoulli distributed random variable, where k is a positive integer, will always be p. As proof,

$E(X^k) = 0^k * q + 1^k * p = p$

Example
Suppose X~Bernoulli(p = 0.78). Find
(a.) E(X) (b.) Var(X) (c.) second moment (d.) third moment
(e.) third central moment.

Solutions:
(a.) $E(X) = p = 0.78$
(b.) $Var(X) = pq = 0.78 * 0.22 = 0.1716$
(c.) $E(X^2) = Var(X) + E(X)^2 = 0.1716 + 0.78^2 = 0.78$

OR
$$\psi'(t) = pe^t \rightarrow \psi''(t) = pe^t \rightarrow \psi''(0) = pe^0 = p = 0.78$$

(d.) $\psi^{(3)}(t) = pe^t \rightarrow \psi^{(3)}(0) = p = 0.78 = E(X^3)$

In fact, all the moments for the Bernoulli distribution will be p.

(e.) $E((X - E(X))^3) = E((X - 0.78)^3) = E((X - 0.78)(X - 0.78)(X - 0.78))$
$= E((X^2 - 1.56X + 0.6084)(X - 0.78))$
$= E(X^3 - 1.56X^2 + 0.6084X - 0.78X^2 + 1.2168X - 0.4746)$
$= E(X^3 - 2.34X^2 + 1.8252X - 0.4746)$
$= E(X^3) - 2.34E(X^2) + 1.8252E(X) - 0.4746$
$= 0.78 - 2.34*0.78 + 1.8252*0.78 - 0.4746$
$= -0.096144$

Binomial Distribution

Recall that a binomial process is really a series of n Bernoulli trials. Each trial has probability p of success. The value of a binomially distributed random variable is the quantity of successes. Since a binomially distributed random variable can only assume integer values between 0 and n, inclusive, the variable is discrete.

Example
Derive the expected value and variance of the binomial distribution.

Solution:
Recall that a binomial process consists of a sequence of Bernoulli trials.
The random variable X with a binomial distribution can be defined as
$X = X_1 + X_2 + ... + X_n$, where each X_i is the resulting value of a particular Bernoulli trial.

$$E(X) = E(X_1 + X_2 + ... + X_n) = E(X_1) + E(X_2) + ... + E(X_n) = p + p + ... + p = np$$

All the Bernoulli trials are independent, so we can use the formula below for the variance of a sum of independent random variables:

$$Var(X) = Var(X_1 + X_2 + ... + X_n) = Var(X_1) + Var(X_2) + ... + Var(X_n)$$
$$= pq + pq + ... + pq = npq$$

Example
Derive the mgf of the binomial distribution.

Solution:
We can again utilize the fact that the X_i's are independent Bernoulli trials.

$$\psi(t) = E(e^{tX}) = E(e^{\wedge}t(X_1 + X_2 + ... + X_n)) = E(e^{tX_1 + tX_2 ... + tX_n})$$
$$= E(e^{tX_1} * e^{tX_2} * ... * e^{tX_n}) = E(e^{tX_1}) * E(e^{tX_2}) * ... * E(e^{tX_n})$$
$$= (pe^t + q) * (pe^t + q) * ... * (pe^t + q)$$
$$= (pe^t + q)^n$$

Formulas for the Binomial Distribution

$E(X) = np$

$Var(X) = npq$

$\psi(t) = (pe^t + q)^n$

Two binomial distributions with the same parameter p but possibly different values for n can be merged into one large binomial distribution. If $X_1 \sim$ Binomial(n_1, p) and $X_2 \sim$ Binomial(n_2, p), and X_1 and X_2 are independent, then $Y = (X_1 + X_2) \sim$ Binomial($n_1 + n_2$, p). The variable Y now represents the number of successes obtained from ($n_1 + n_2$) trials, each of which has probability p of succeeding. We can discover Y's distribution by building its moment-generating function:

Let the mgf of X_1 be $\psi_1(t) = \left(pe^t + q\right)^{n_1}$

Let the mgf of X_2 be $\psi_2(t) = \left(pe^t + q\right)^{n_2}$

Since X_1 and X_2 are independent, the mgf of $Y = (X_1 + X_2)$ can be found by multiplying the mgf's of X_1 and X_2:

$$\psi_Y(t) = \left(pe^t + q\right)^{n_1} * \left(pe^t + q\right)^{n_2} = \left(pe^t + q\right)^{n_1 + n_2}$$

$\psi_Y(t)$ represents the mgf for a binomial distribution with parameters p and $(n_1 + n_2)$, so Y must have this distribution.

Example
The probability that a computer virus infects a server is 4.5%. A hacker attempts to upload the virus to 500 servers. All the servers are independent from each other.

(a.) What is the expected number of servers that will accept the virus?
(b.) What is the variance in the number of servers taking the virus?
(c.) What is the standard deviation for the amount of servers becoming infected?

Solutions:
Let n = 500, p = 0.045, q = 0.955

(a.) $E(X) = 500 * 0.045 = 22.5$
(b.) $Var(X) = 500 * 0.045 * 0.955 = 21.4875$
(c.) $s = \sqrt{Var(X)} = \sqrt{21.4875} = 4.635$

Example
A lottery designed a scratch-off game so that a given customer has a 7.8% chance of winning. If 1,000 customers buy the game, what is the expected number of winners and the standard deviation in the amount of winners?

Solution:
Let X = number of winners out of 1,000
Let n = 1,000, p = 0.078, and q = 0.922

$E(X) = 1,000 * 0.078 = 78$
$Var(X) = s^2 = npq = 1,000 * 0.078 * 0.922 = 71.916$ → $s = \sqrt{71.916} = 8.48$

Example
Suppose X has a binomial distribution.
You are given that E(X) = 176 and Var(X) = 21.12. What are n and p?

Solution:
$E(X) = np = 176$ $Var(X) = npq = 21.12$
                                  ↓
                              $176*q = 21.12$
                              $q = 0.12$ → $p = 0.88$

np = 176
n * 0.88 = 176
n = 200

Example
The random variable X has a binomial distribution with 25 trials and chance of success 45% per trial. Find the chance that X differs from its expected value by 2 or more successes.

Solution:
$E(X) = np = 25 * 0.45 = 11.25$

$$\Pr(|X - E(X)| \geq 2) = 1 - (f(10) + f(11) + f(12) + f(13))$$

$$f(10) = \binom{25}{10} * 0.45^{10} * 0.55^{15} = 0.14189$$

$$f(11) = \binom{25}{11} * 0.45^{11} * 0.55^{14} = 0.15831$$

$$f(12) = \binom{25}{12} * 0.45^{12} * 0.55^{13} = 0.15111$$

$$f(13) = \binom{25}{13} * 0.45^{13} * 0.55^{12} = 0.12364$$

$$\Pr(|X - E(X)| \geq 2) = 0.42505$$

Section 6-2: Poisson Distribution

The random variable in a Poisson distribution usually represents a count of occurrences of some phenomenon over a fixed time period. X can assume only non-negative integers. That is, X can take integer values from 0 (inclusive) up to positive infinity. X might represent the number of customers arriving at a store over an hour, the number of calls to customer support, or the number of car accidents on a highway. The Poisson distribution has a single parameter called λ which represents the rate of occurrences of the phenomenon per unit of time. The parameter λ can be any real value greater than 0. Lambda cannot be negative because a negative amount of occurrences in a time period is impossible. Both the expected value and variance of X equal λ.

The Poisson distribution results from the binomial distribution as $n \to \infty$ and $p \to 0$. Over a time interval of nonzero width, there exist infinitely many moment of time. In other words, the number of "trials" is infinitely big. At each moment, a success could occur with probability p. The probability of a success at a moment will be very, very small. However, over a large time interval like a day, the process might encounter many successes. If you are faced with a binomial distribution with n being large and p being quite small, you can approximate the distribution with a Poisson distribution setting the parameter to $\lambda = E(X) = np$. Typically, n should be above 100 to approximate a binomial distribution with a Poisson distribution.

The pf for a Poisson distributed random variable appears below. The variable X can take values between 0, inclusive, and $+\infty$, exclusive. X represents the number of occurrences of some phenomenon.

Probability Function for the Poisson Distribution

$$f(x) = e^{-\lambda} \cdot \frac{\lambda^x}{x!}, \text{ for x a non-negative integer}$$

The probability function for the Poisson distribution generates a set of discrete points which form the outline of a skewed bell curve. The probability function generally peaks around $x = \lambda$. The function has a long tail on the right side which quickly approaches the x-axis. The pf will still produce positive probabilities for large x-values, although these probabilities will be close to 0.

The distribution function for the Poisson distribution is a step function. Each step has an x-length of 1 unit. At every integer x-value greater than or equal to 0, the distribution function jumps upward by f(x).

Plots of the pf and df for various values of λ appear below. In the case of the pf's, the probability values should be zero at all x-values which are not positive integers. However, plotting a sequence of open circles and segments along the x-axis complicates the pf graph.

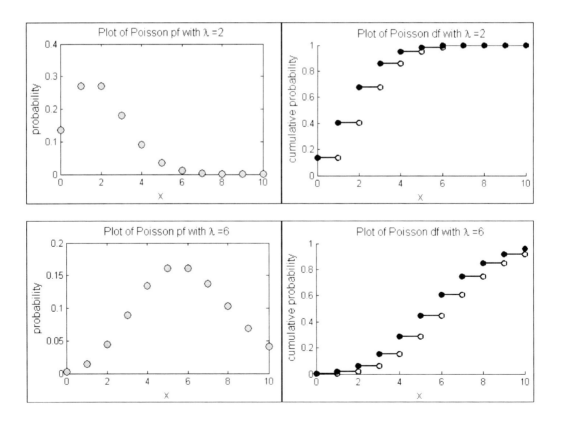

Example
TP Oil Industries experiences many catastrophic oil spills in its regular operations. The firm believes that the quantity of spills in a year is Poisson distributed. Over the course of a year, it expects 6 spills.
(a.) What is the chance the company has no spills this year?
(b.) Find the probability the firm has more than 5 spills this year?

Solutions:

(a.) $\Pr(X = 0) = e^{-6} \cdot \dfrac{6^0}{0!} = e^{-6} = 0.002479$

(b.) Rather than directly computing f(6) + f(7) + f(8) + ..., you should compute the chance that $0 \leq X \leq 5$ and then subtract this probability from 1.

f(0) = 0.002479, f(1) = 0.01487, f(2) = 0.04462, f(3) = 0.08924,
f(4) = 0.13385, f(5) = 0.16062

$\Pr(0 \leq X \leq 5) = 0.4457$

probability more than 5 spills = $\Pr(X \geq 5) = 0.5543$

Formulas for the Poisson Distribution

$E(X) = \lambda$

$Var(X) = \lambda$

$\psi(t) = e^{\lambda(e^t - 1)}$

The mode of a Poisson distribution occurs fairly close to the value of λ. If λ is a positive integer, then the distribution has two modes occurring at λ and $\lambda - 1$. On the other hand, if λ is a floating point value, then the distribution has just one mode, which arises at the greatest integer just less than λ.

Example
Juan owns a taco stand that receives customers according to a Poisson process. Juan expects 5 customers per hour.
(a.) What is the variance in the number of customers coming in one hour?
(b.) What is the chance that Juan serves exactly 4 customers in the next hour?
(c.) What is the chance that no customers come to Juan's stand in the next hour?

> (d.) What is the probability that Juan receives more than 2 customers in the next hour?
> (e.) Find all the modes of the distribution.
>
> Solutions:
> Let X = # of customers arriving in 1 hour. X ~ Poisson($\lambda = 5$).
> (a.) Var(X) = λ = 5
>
> (b.) Pr(4 customers) = f(4) = $e^{-5} \cdot \dfrac{5^4}{4!} \approx 0.17547$
>
> (c.) Pr(0 customers) = f(0) = $e^{-5} \cdot \dfrac{5^0}{0!} = e^{-5} \approx 0.006738$
>
> (d.) Pr(X > 2) = Pr($X \geq 3$) = 1 − (f(0) + f(1) + f(2))
>
> $= 1 - \left(e^{-5} \cdot \left(\dfrac{5^0}{0!} + \dfrac{5^1}{1!} + \dfrac{5^2}{2!} \right) \right) = 1 - \left(e^{-5} \cdot (1 + 5 + 12.5) \right) = 1 - \left(e^{-5} \cdot 18.5 \right)$
>
> ≈ 0.87535
>
> (e.) Since λ is a positive integer, the distribution has two modes. The modes are 5 and 4.

The Poisson distribution can be expanded to various intervals of time. Assume that X~Poisson(λ). For each unit of time, we expect λ occurrences of the phenomenon of interest. For half a unit of time, we would expect 0.5λ occurrences. If we monitored an unchanging situation for 10 units of time, we would expect 10λ happenings. In general, if we know that the distribution for the number of occurrences, X, over one time unit is Poisson(λ), and Y represents the number of occurrences over a time interval of t units, then Y~Poisson($t\lambda$). Independence exists between time intervals so that the quantity of occurrences in one interval has no influence over the number of happenings in another interval. In fact, the full history of previous occurrences can have no impact on the quantity of future occurrences.

In general, the sum of several independent random variables which are each Poisson distributed will also be Poisson distributed. To find the parameter for the sum, simply add up all the parameters for the constituent distributions.

Sum of Poisson distributed Random Variables
If $X_1, X_2, \ldots, X_n$ are independent random variables, and X_i~Poisson(λ_i), then the sum $Y = X_1 + X_2 + \ldots + X_n$ is a Poisson distributed random variable with parameter $\lambda = \lambda_1 + \lambda_2 + \ldots + \lambda_n$.

Example
A steel mill produces bars of steel of various lengths. The manager believes that the quantity of defects per 5 feet of steel is Poisson distributed with variance equal to 1.
(a.) What is the distribution for the number of defects in 30 feet of steel?

(b.) What is the probability that a bar of steel 50 feet long has 11 or 12 defects?
(c.) What is the probability that a 2-foot bar of steel has at least one defect?

Solutions:
Let X = number of defects per 5 feet of steel
X~Poisson($\lambda = 1$)

(a.) Let Y = number of defects per 30 feet of steel
Y ~ Poisson($\lambda = 6*1 = 6$)

(b.) Let Y = number of defects per 50 feet of steel
Y ~ Poisson($\lambda = 10$)

$$\Pr(Y = 11 \text{ or } Y = 12) = f(11) + f(12) = e^{-10} \cdot \frac{10^{11}}{11!} + e^{-10} \cdot \frac{10^{12}}{12!} \approx 0.20852$$

(c.) Let Y = number of defects per 2 feet of steel
Y ~ Poisson($\lambda = 0.4*1 = 0.4$)

$$\Pr(Y \geq 1) = 1 - \Pr(Y = 0) = 1 - e^{-0.4} * \frac{0.4^0}{0!} \approx 1 - 0.67032 \approx 0.32968$$

Example
Suppose that the number of homicides that occur each day in the city of Murderapolis is Poisson distributed with an expected value of 2. What is the probability that between 11 and 15 murders, inclusive, occur in the city in the upcoming week?

Solution:
X = # of murders in a single day
X ~ Poisson($\lambda = 2$).

Y = # of murders that occur over 7 days
Since t = 7 and the original $\lambda = 2$,
Y ~ Poisson($t\lambda = 14$)

$$\Pr(11 \leq Y \leq 15) = f(11) + f(12) + f(13) + f(14) + f(15)$$
$$= e^{-14} \cdot \left(\frac{14^{11}}{11!} + \frac{14^{12}}{12!} + \frac{14^{13}}{13!} + \frac{14^{14}}{14!} + \frac{14^{15}}{15!} \right) \approx 0.49368$$

Example
The customer support department for an online company receives emails from customers. The quantity of emails received per hour appears to be Poisson distributed. However, the parameter for the distribution seems to vary depending on the time of day. Between the hours of 12 AM and 7 AM, the corporation expects 4 emails per hour. From 7 AM to 6

PM, the company expects 9 emails per hour. From 6 PM to 12 AM the next day, the company expects 5 emails per hour. Assume that all emails arrive independently of each other. Find the expected value and variance of the number of emails received in a day.

Solution:
Let X = # emails received from 12 AM to 7 AM
Let Y = # emails received from 7 AM to 6 PM
Let Z = # emails received from 6 PM to 12 AM
Let T = (X + Y + Z) = total # emails received in a day

$X \sim$ Poisson($\lambda_X = 28$)
$Y \sim$ Poisson($\lambda_Y = 99$)
$Z \sim$ Poisson($\lambda_Z = 30$)
$T \sim$ Poisson($\lambda_T = 157$)

$E(T) = 157$, $Var(T) = 157$

Example
A computer hacker is attempting to break into 500 random computers around the world. Assume that all intrusion attempts are independent from each other. The hacker computes that the chance of a successful intrusion on any given computer is 0.1%. Find the chance that he breaks into less than 2 systems using…
(a.) the precise binomial distribution.
(b.) the Poisson distribution approximation.

Solutions:
(a.) $\Pr(X = 0 \cup X = 1) = f(0) + f(1) = 0.60638 + 0.30349 = 0.90987$
(b.) $\lambda = np = 500 * 0.001 = 0.5$

$\Pr(X = 0 \cup X = 1) = f(0) + f(1) = 0.60653 + 0.30327 = 0.90980$

Section 6-3: Negative Binomial and Geometric Distributions

The negative binomial distribution is a higher abstraction of the regular binomial distribution. The experiment consists of a number of trials, each of which has a Bernoulli distribution with probability of success p. Each trial could be a success or failure. Additionally, all the trials are independent. In the case of the negative binomial distribution, the number of trials is not fixed. Instead, the experiment ends as soon as the rth success is observed. It is possible that the rth success is never obtained in a finite number of trials. The negative binomial distribution has 2 parameters, r and p.

The random variable X which has a negative binomial distribution represents the number of failures until obtaining the rth success. X does not count successes occurring. If the experiment produces a string of r consecutive successes from the start, then X would be 0. Typically, the experiment consists of successes intermixed with failures. As stated earlier, the experiment could consists of thousands of trials before obtaining the rth success despite a high p-value, in which case X will be quite large. X can only assume non-negative integers.

Example
The string of successes (1's) and failures (0's) from an experiment is shown below:
0, 0, 1, 0, 1, 1, 1, 0, 0, 0, 1, 0, 1, 1, 0, 0, 0, 0, 1

Compute the value of x from a negative binomial distribution when r equals...
(a.) r = 1   (b.) r = 2   (c.) r = 5   (d.) r = 8

Solutions:
(a.) X(1) = 2   (b.) X(2) = 3   (c.) X(5) = 6   (d.) X(8) = 11

Key Notation for the Negative Binomial Distribution
p = probability of success
q = probability of failure = (1 – p)
r = # of successes needed
X = # of failures before the rth success
(X + r) = total # of trials
(X + r – 1) = # of trials in which the failures can occur

The discrete probability function for a random variable with a negative binomial distribution is shown below. The formula somewhat resembles that for a regular binomial distribution. It includes a combination term, a factor for successes, and a factor for failures.

Probability Function for the Negative Binomial Distribution

$$f(x) = \binom{x+r-1}{x} \bullet p^r \bullet q^x$$

The term $\binom{x+r-1}{x}$ shows the number of ways to pick x failures from among the first $(x + r – 1)$ trials. Note that the $(x + r)^{th}$ trial must be a success, so it is not considered in the combination. The combination is needed because there are multiple ways to position x failures in the string of trials. The factor p^r gives the probability of r successes happening. The factor q^x provides the chance that x failures happen. Several plots of the negative binomial distribution's pf and df under different parameters are displayed here. Note that the negative binomial's pf resembles a skew symmetric bell curve.

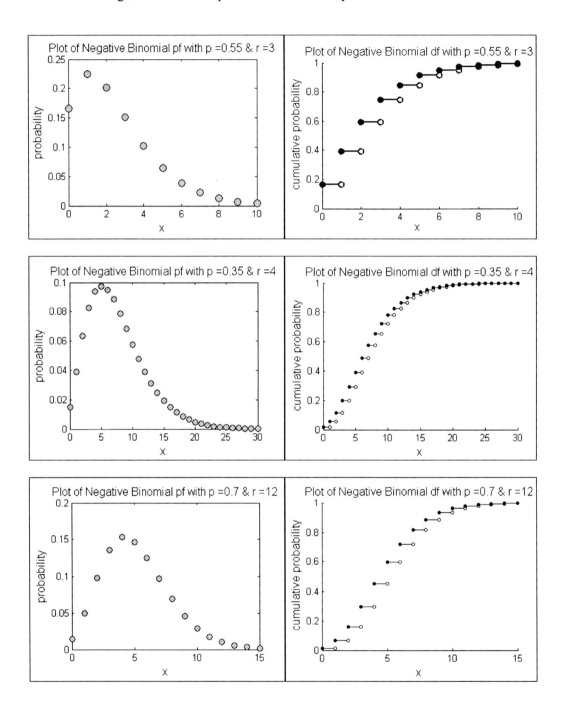

Formulas for the Negative Binomial Distribution

$$E(X) = \frac{rq}{p}$$

$$Var(X) = \frac{rq}{p^2}$$

$$\psi(t) = \left(\frac{p}{1-qe^t}\right)^r$$

Example
Jamal is shooting hoops. The chance that he makes any given shot is 0.6. All the shots are independent from each other, and he shoots from the same position on the floor.
(a.) What is the expected number of misses before Jamal makes his 7th basket?
(b.) What is the expected total number of shots (failures and successes) required for Jamal to make 7 baskets?

Solutions:
(a.) $X \sim $ NegBinomial(r = 7, p = 0.6)
$$E(X) = \frac{7*0.4}{0.6} = 4.6667 \approx 5$$

(b.) E(total number of shots for 7 baskets) = E(X + r) = E(X) + E(r) = E(X) + r
$$= 4.6667 + 7 = 11.6667 \approx 12.$$

Example
Schmidt has a biased coin that he repeatedly flips. He knows that the expected number of tails before obtaining 4 heads is 15.
(a.) Find p, the probability of obtaining heads on any given flip.
(b.) Compute the chance that Schmidt encounters fewer than 3 tails before obtaining 2 total heads.
(c.) What is the variance in the number of tails obtained before getting 8 heads?
(d.) Compute the probability that Schmidt must toss the coin exactly 20 times to get precisely 5 heads.

Solutions:
(a.) Let X = # tails before obtaining r heads

$$E(X) = 15 = \frac{rq}{p} \to 15 = \frac{4(1-p)}{p} \to 15p = 4 - 4p \to 19p = 4 \to p = 4/19$$

(b.) r = 2
$\Pr(X < 3) = \Pr(X \leq 2) = f(0) + f(1) + f(2)$

$$f(0) = \binom{0+2-1}{0} \bullet p^2 \bullet q^0 = 1 \bullet p^2 \bullet 1 = p^2 = 0.04432$$

$$f(1) = \binom{1+2-1}{1} \bullet p^2 \bullet q^1 = \binom{2}{1} \bullet p^2 \bullet q^1 = 2p^2 q = 0.06998$$

$$f(2) = \binom{2+2-1}{2} \bullet p^2 \bullet q^2 = \binom{3}{2} \bullet p^2 \bullet q^2 = 3p^2 q^2 = 0.08287$$

$\Pr(X < 3) = 0.04432 + 0.06998 + 0.08287 = 0.19717$

(c.)
$$Var(X) = \frac{rq}{p^2} = \frac{8*(1-4/19)}{(4/19)^2} = 142.5$$

(d.) Out of the 20 tosses, 5 tosses are heads and 15 are tails.
$$f(15) = \binom{15+5-1}{15} \bullet p^5 \bullet q^{15} = \binom{19}{15} \bullet p^5 \bullet q^{15} = 3876 p^5 q^{15} = 0.04624$$

Example
Ezra has a water balloon launcher and is catapulting water balloons at his friend's house. He knows that the chance that a particular balloon hits its target is 32%. He must make 8 hits for the mission to be successful.
(a.) Calculate the expected number of misses before the mission is over.
(b.) Calculate the expected total number of balloons used (hits and misses) before the mission is complete.
(c.) What is the variance in the number of misses before finishing the mission?
(d.) What is the chance that he requires 13, 14, or 15 balloons to complete the mission?
(e.) What is the probability that the number of misses is 9 or 10?

Solutions:
(a.) Let X = # misses; X ~ NegBinomial(p = 0.32, r = 8)

$$E(X) = \frac{rq}{p} = \frac{8*0.68}{0.32} = 17$$

(b.) $E(X+r) = E(X) + r = 17 + 8 = 25$

(c.) $Var(X) = \dfrac{rq}{p^2} = \dfrac{8*0.68}{0.32^2} = 53.125$

(d.) 13 total balloons entails 5 misses
14 total balloons entails 6 misses
15 total balloons entails 7 misses

$\Pr = f(5) + f(6) + f(7)$

$f(5) = \binom{5+8-1}{5} \cdot p^8 \cdot q^5 = \binom{12}{5} \cdot 0.32^8 \cdot 0.68^5 = 792 \cdot 0.32^8 \cdot 0.68^5 = 0.01266$

$f(6) = \binom{6+8-1}{6} \cdot p^8 \cdot q^6 = \binom{13}{6} \cdot 0.32^8 \cdot 0.68^6 = 1716 \cdot 0.32^8 \cdot 0.68^6 = 0.01865$

$f(7) = \binom{7+8-1}{7} \cdot p^8 \cdot q^7 = \binom{14}{7} \cdot 0.32^8 \cdot 0.68^7 = 3432 \cdot 0.32^8 \cdot 0.68^7 = 0.02537$

$\Pr = 0.01266 + 0.01865 + 0.02537 = 0.05668$

(e.)
$\Pr = f(9) + f(10) = \binom{9+8-1}{9} \cdot p^8 \cdot q^9 + \binom{10+8-1}{10} \cdot p^8 \cdot q^{10}$

$= \binom{16}{9} \cdot 0.32^8 \cdot 0.68^9 + \binom{17}{10} \cdot 0.32^8 \cdot 0.68^{10} = 0.03910 + 0.04520 = 0.08430$

Geometric Distribution

The geometric distribution is a special case of the negative binomial distribution with the parameter r = 1. A random variable with a geometric distribution has a value equaling the quantity of failures until the first success. Once you observe the first success, the process halts. The geometric distribution has just one parameter, which is p. The string of outcomes for the individual trials would appear as: 0, 0, 0, ..., 0, 0, 1.

The geometric distribution is the only discrete distribution with the memoryless property. The string of previous failures has no bearing on whether the next outcome will be a success. If we observed 100 failures and are still waiting for the first success, then the probability of success on the next trial is still p. The independence between all trials contributes to the memoryless nature of the process. The entire process essentially restarts at every trial.

The probability function for the geometric distribution is a simplified version of that for the negative binomial distribution. The combination is absent because there is only one way to organize the failures among all the trials; the first x trials will all be failures. You simply need to multiply the chance of getting x failures with the chance of getting one success. As observed in the following graphs, the pf values tend to follow an invisible curve with the greatest value being f(0) and diminishing thereafter. The value

$X = 0$ is observed when the very first trial is a success (and there are no failures). The random variable can take integer values greater than or equal to 0.

Probability Function for the Geometric Distribution

$$f(x) = \begin{cases} p \bullet q^x, & \text{if x is a nonnegative integer} \\ 0, & \text{else} \end{cases}$$

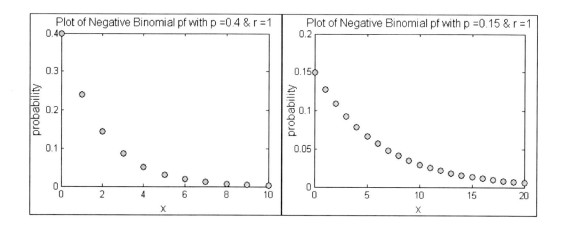

Formulas for the Geometric Distribution

$$E(X) = \frac{q}{p}$$

$$Var(X) = \frac{q}{p^2}$$

$$\psi(t) = \frac{p}{1 - qe^t}$$

Example
A pilot is trying to land on a runway. The chance that the plane successfully touches the runway is 70%. If he misses the runway, then he must circle the plane around and attempt to land again. The pilot will continue landing attempts until he is successful.
(a.) What is the expected number of missed landing attempts before succeeding?
(b.) What is the expected total number of landing attempts?
(c.) What is the variance in the number of missed landing attempts?
(d.) What is the chance that he will miss the runway fewer than 5 times?

(e.) What is the probability that he requires exactly 6 attempts before landing correctly?

Solutions:
Let X = # of missed landing attempts; X ~ Geometric(p = 0.7)

(a.) $E(X) = \dfrac{q}{p} = \dfrac{0.3}{0.7} = 3/7$

(b.) $E(X + 1) = E(X) + 1 = 10 / 7$

(c.) $Var(X) = \dfrac{q}{p^2} = \dfrac{0.3}{0.7^2} = \dfrac{0.3}{0.49} = 0.61224$

(d.) $Pr = f(0) + f(1) + f(2) + f(3) + f(4)$
$= p(q^0 + q^1 + q^2 + q^3 + q^4)$
$= 0.7(0.3^0 + 0.3^1 + 0.3^2 + 0.3^3 + 0.3^4)$
$= 0.7 * 1.4251$
$= 0.99757$

(e.) $f(5) = pq^5 = 0.7 * 0.3^5 = 0.001701$

Example
Phong is a terrible driver but is attempting to get his driver's license. He will take the driver's test until he passes. Assume that Phong takes each test under the same conditions. The chance that he passes any given test is 6.5%. What is the chance that he fails the test between 6 and 9 times, inclusive, before passing?

Solution:
Let X = # of failed tests; X ~ Geometric(p = 0.065)

$Pr(6 \leq X \leq 9) = f(6) + f(7) + f(8) + f(9) = p(q^6 + q^7 + q^8 + q^9)$
$= 0.065(0.935^6 + 0.935^7 + 0.935^8 + 0.935^9) = 0.1575$

Example
Enoch is a compulsive gambler playing scratch-off games. He tells himself that he will stop gambling as soon as he wins the jackpot. The chance of any game producing the jackpot is 0.25%. Compute the following:
(a.) Expected total number of games played in order to win
(b.) Variance in the total number of games played in order to win
(c.) Chance Enoch wins with the 7th game

Solutions:
X = number of lost games played

Digital Actuarial Resources   *Comprehensive Probability Review for Actuarial Exams*

$(X + 1)$ = total number of games played (including the single win)

$X \sim \text{Geometric}(p = 0.0025)$

(a.) $E(X + 1) = E(X) + 1 = \dfrac{0.9975}{0.0025} + 1 = 400$

(b.) $Var(X + 1) = Var(X) = \dfrac{0.9975}{0.0025^2} = 159{,}600$

(c.) $f(7) = 0.0025 * 0.9975^7 = 0.002457$

Section 6-4: Hypergeometric Distribution

The hypergeometric distribution is a unique discrete distribution in which trials are dependent upon each other. A hypergeometric process involves a relatively small pool of objects, each of which can belong to one of two categories. A success occurs when an object of type 1 is drawn. A failure arises when the drawn object is of type 2. Once the experimenter uses an object from the pool, he does not replace it. The "without replacement" nature of the process means that each draw is dependent upon all the previous draws. If the experimenter drew 5 objects of type 1 in the first 5 trials, then the chance that the sixth object is of type 1 will be diminished because there are fewer objects of type 1 remaining in the pool. The hypergeometric distribution has three parameters—A, B, and n—which are described below:

Key Notation for the Hypergeometric Distribution
A = # of available objects at the start in condition type 1
B = # of available objects at the start in condition type 2
T = (A + B) = total # of available objects at the start
n = # objects drawn
X = # of objects in condition type 1 actually drawn
(n − X) = # of objects in condition type 2 actually drawn

Initial Proportions of Objects
percentage of objects of type 1 = $p = \dfrac{A}{T}$

percentage of objects of type 2 = $q = \dfrac{B}{T}$

The probability function has a unique domain of x-values over which it is positive. The random variable must be a non-negative integer. The experimenter selects n objects. The value of n must be constrained to lie between 1 and T, inclusive. In other words, he will draw at least 1 object, but he cannot draw more objects than are available. If n exceeds B, then he is guaranteed to obtain at least $(n - B)$ objects in class 1. Likewise, when n exceeds A, then the experimenter must retrieve at least $(n - A)$ objects in class 2. In addition, it is impossible to draw more objects from class 1 than actually exist in the pool. That is, X cannot exceed A. The bounds on X are:

$$\max(n - B, 0) \leq X \leq \min(A, n)$$

The hypergeometric pf is developed with the help of principles from the simple sample space distribution. To find the probability that $X = x$, we need to find the number of outcomes in which $X = x$ and divide by the total number of outcomes in the experiment. We must be consistent in how we count elements in the numerator and denominator. Let's assume that order doesn't matter. The total number of ways of selecting n elements from a pool of $(A + B)$ elements is $_{A+B}C_n$. The number of ways of drawing objects such that x objects of type 1 and $(n - x)$ objects of type 2 are selected must be $_AC_x \cdot _BC_{n-x}$.

Probability Function for the Hypergeometric Distribution

$$f(x) = \begin{cases} \dfrac{\binom{A}{x} \cdot \binom{B}{n-x}}{\binom{A+B}{n}}, & \text{if } \max(n - B, 0) \leq X \leq \min(A, n) \\ 0, & \text{else} \end{cases}$$

Example
Suppose a hat contains 8 green cards and 6 blue cards. Ezekiel must draw 4 cards from the hat. He will not replace each card as he draws it. Let X = number of greed cards selected.

(a.) What are the bounds on X?
(b.) What is the initial probability of obtaining a green card and the initial probability of a blue card?
(c.) If the first two draws are blue cards, what is the chance the third card is green?

Solutions:
X ~ Hypergeometric(A = 8, B = 6, n = 4)

(a.)

$\max(4-6, 0) \leq X \leq \min(8, 4) \rightarrow 0 \leq X \leq 4$

(b.)
initial probability green = 8 / 14
initial probability blue = 6 / 14

(c.)
Pr(third is green | first is blue and second is blue) = 8 / 12

Example
DeAndre is assembling a football team and needs players. He will select from youths in two districts: the north side and the south side. Assume he picks players randomly and without favoritism. The north side has 20 kids, and the south side has 8 kids. He must select 11 kids to form a team.

(a.) What is the chance that DeAndre's team has all 8 kids from the south side?
(b.) Find the probability that the team has precisely 6 or 7 kids from the north side.
(c.) Compute the chance that the team has fewer than 5 kids from the north side.

Solutions:
Let X = number of chosen kids which are from the north side
X ~ Hypergeometric(A = 20, B = 8, n = 11)

(a.)
Pr(8 kids from south side) = Pr(3 kids from north side) = f(3) =

$$= \frac{\binom{20}{3} \cdot \binom{8}{11-3}}{\binom{28}{11}} = \frac{1140 * 1}{21{,}474{,}180} = 0.000053087$$

(b.)
$$f(6) + f(7) = \frac{1}{\binom{28}{11}} * \left(\binom{20}{6} \cdot \binom{8}{11-6} + \binom{20}{7} \cdot \binom{8}{11-7} \right)$$

$$= \frac{1}{\binom{28}{11}} * (38{,}760 \bullet 56 + 77{,}520 \bullet 70) = 0.35377$$

(c.)
The value X has zero probability when X falls below 3. For instance, in the impossible case that $X = 2$, exactly 2 kids must come from the north side and 9 kids must come from the south side. However, the south side only has 8 kids. Therefore, $Pr(X = 2) = 0$.

We only need to consider the cases when X = 3 or 4.

$$f(3) + f(4) = \frac{1}{\binom{28}{11}} * \left(\binom{20}{3} \bullet \binom{8}{11-3} + \binom{20}{4} \bullet \binom{8}{11-4} \right)$$

$$= \frac{1}{\binom{28}{11}} * (1{,}140 \bullet 1 + 4{,}845 \bullet 8) = 0.001858$$

If the sizes of A and B are large relative to x and n, then you could approximate the hypergeometric distribution with the binomial distribution. The binomial process is "with replacement" because the probability of success is constant for all trials; past outcomes do not affect future outcomes. When A and B are relatively large, the fact that we are not replacing drawn objects has little impact on the future probability of drawing an object of type 1.

Example
A jar has 35 black and 45 red chips. Erwin selects 8 chips from the jar. Find the chance that he obtains 2, 3, or 4 black chips assuming…

(a.) Erwin does not replace the drawn chips, and X = (number of black chips) has a hypergeometric distribution.
(b.) Erwin replaces each drawn chip, and X has a binomial distribution.

Solutions:
(a.) A = 35, B = 45, n = 8

$$f(2) + f(3) + f(4) = \frac{1}{\binom{35+45}{8}} * \left(\binom{35}{2} \bullet \binom{45}{8-2} + \binom{35}{3} \bullet \binom{45}{8-3} + \binom{35}{4} \bullet \binom{45}{8-4} \right)$$

$$= \frac{1}{2.89875 * 10^{10}} * (595 \bullet 8{,}145{,}060 + 6{,}545 \bullet 1{,}221{,}759 + 52{,}360 \bullet 148{,}995)$$

$$= \frac{2.06441 * 10^{10}}{2.89875 * 10^{10}} = 0.71217$$

(b.) X ~ Binomial(n = 8, p = 35/80 = 0.4375)

$$f(2) + f(3) + f(4) = 0.16977 + 0.26408 + 0.25674 = 0.69059$$

Notice that the binomial approximation is quite close to the precise probability value from the hypergeometric distribution.

Formulas for the Hypergeometric Distribution

$$E(X) = \frac{nA}{T} = np$$

$$Var(X) = \frac{T-n}{T-1} * npq$$

Example
A hat has 50 yellow tokens and 28 red tokens. The experimenter draws 12 tokens and counts the number of yellow tokens. Find the variance in the number of yellow tokens drawn assuming...

(a.) the process is without replacement; X has a hypergeometric distribution
(b.) the process is with replacement; X is binomially distributed

Solutions:
(a.) X = # yellow tokens
   X ~ Hypergeometric(A = 50, B = 28, n = 12)

$$Var(X) = \frac{78-12}{78-1} * 12 * \frac{50}{78} * \frac{28}{78} = \frac{6}{7} * 2.76134 = 2.36686$$

(b.) X ~ Binomial(n = 12, p = 0.64103)

$$Var(X) = 12 * 0.64103 * 0.35897 = 2.7613$$

Observe that the variance calculation under the assumption of "without replacement" is a scaled version of the variance under the assumption of "with replacement." The scale factor is $6/7$.

Section 6-5: Multinomial Distribution

The multinomial distribution is a generalization of the binomial distribution. A random variable with a multinomial distribution consists of a string of n independent trials. Each trial produces an outcome that can fall in one of k categories. The quantity of outcomes existing in the ith category is denoted x_i. The random variable X is really a vector of counts: $\vec{X} = (x_1, x_2, ..., x_k)$. Since the trials are independent, a trial that falls in category j will have no effect on the potential categories for future trials. Each category has a probability p_i of success (the chance that any given trial will yield an outcome for category i). The probabilities of success for the classes form a vector. All the p_i's must sum to 1. In addition, all the elements in $\vec{x}$ must sum to n.

The classes need to be mutually exclusive. An outcome must activate precisely one category. If the classes are traits of people, it must be impossible for a person to simultaneously exist in more than (or less than) one category. An example of an experiment that fails the requirement of mutually exclusive categories arises when the categories are symptoms of schizophrenia. Common symptoms are delusions of grandeur, irrational fantasies, voices in one's head, and paranoia. A schizophrenic patient likely has multiple symptoms, and the symptoms can change spontaneously. In other words, a patient could exist in a few categories.

Parameters for the Multinomial Distribution
n = total number of trials
$\vec{p} = (p_1, p_2, ..., p_k)$

The multinomial distribution is a multivariate distribution. The number of outcomes lying in category i forms the random variable X_i. The random variables X_i and X_j are dependent upon each other since a success in one category necessarily implies failure in all other categories. Additionally, any two distinct variables for category counts have a negative covariance and negative correlation.

The probability function resembles that for the binomial distribution. The statement that $X_1 = x_1, X_2 = x_2, ..., X_k = x_k$ means that x_1 of the trials have to fall in category 1, x_2 of the trials must fall in class 2, and so on. Since we know the probability of success and the number of successes in each category, and we know that all the trials are independent, we can multiply the probabilities together, forming $p_1^{x_1} p_2^{x_2} ... p_k^{x_k}$. The multinomial coefficient is employed because there are many ways to choose the successes for each category from among all n trials.

Probability Function for the Multinomial Distribution
$$f(x_1, x_2, ..., x_k) = \binom{n}{x_1, x_2, ..., x_k} \cdot p_1^{x_1} p_2^{x_2} ... p_k^{x_k}$$

The marginal distribution of each random variable X_i is a binomial distribution with parameters n and p_i. Every trial will either produce an outcome lying in category i or outside category i. When the trial increments the count for X_i, a "success" has occurred. Otherwise, a failure occurs. Several traits for the binomial distribution of X_i include:

$$E(X_i) = np_i$$
$$Var(X_i) = np_i(1-p_i)$$

A subset of the X_i random variables, when summed together, will also have a binomial distribution. Consider the case of 2 random variables—X_i, which represents the number of outcomes in category i, and X_j, which stores the number of outcomes in category j. Assume the number of categories exceeds 2. Every trial will either yield an outcome in (category i ∪ category j) or an outcome outside the categories of interest. A "success" occurs when the outcome is in (category i ∪ category j). The probability of success is $(p_i + p_j)$. The number of trials will still be n.

Illustration of the Binomial Distribution as a Special Type of Multinomial Distribution

To show that the binomial distribution is a specific type of multinomial distribution, assume that 2 categories exist, which are labeled success and failure. The categorical random variables are:

X_1 = number of successes
X_2 = number of failures = $n - X_1$

The chance of success is $p_1 = p$, and the chance of failure is $p_2 = 1 - p$. Since X_2 and p_2 are dependent upon X_1 and p_1, respectively, we really have just one categorical variable named X_1 (or simply X). We are only concerned about the quantity of successes. Any trial which is not a success must be a failure. The probability function becomes

$$f(x_1, x_2) = \binom{n}{x_1, x_2} \bullet p_1^{x_1} p_2^{x_2} = \binom{n}{x_1, n-x_1} \bullet p^{x_1}(1-p)^{n-x_1}$$
$$= \frac{n!}{x_1!(n-x_1)!} \bullet p^{x_1}(1-p)^{n-x_1}$$

Let $X = X_1$. Then,

$$f(x) = \frac{n!}{x!(n-x)!} \bullet p^x(1-p)^{n-x} = \binom{n}{x} \bullet p^x q^{n-x}$$

The last expression is the probability function for a binomial distribution.

Example
A worker in the human resources department at a large corporation is processing applications for employment. The worker randomly places each application in one of three bins: "shred," "request phone interview," and "claim job is filled." The probability

for each category, respectively, is: 0.62, 0.33, 0.05. The HR worker receives 8 applications.

(a.) What is the chance that, among the 8 applications, the HR worker shreds 3 of them, requests a phone interview with 4 applicants, and tells 1 applicant the job is filled?
(b.) Find the variance of the distribution for the new random variable ((number of shredded applications) + (number of applicants told job is filled)).

Solutions:
(a.)
The random variable is $\vec{X} = (x_1, x_2, x_3)$, where
X_1 = number of shredded applications
X_2 = number of applications with phone interview
X_3 = number of applicants told job is filled

In this example, n = number of applications = 8.
Additionally, $\vec{p} = (0.62, 0.33, 0.05)$.
We want to find the probability of $\vec{X} = (3, 4, 1)$.

$$f(3, 4, 1) = \binom{8}{3, 4, 1} \bullet 0.62^3 \bullet 0.33^4 \bullet 0.05^1 = 280 \bullet 0.0001413 = 0.03957$$

(b.)
The new random variable $(X_1 + X_3)$ has a binomial distribution. A "success" occurs when the outcome of a trial lies either in category 1 or category 3. A "failure" arises when the outcome rests in the other category (the applicant has a phone interview). The chance of success is (0.62 + 0.05). The number of trials remains n = 8.

$(X_1 + X_3) \sim$ Binomial(n = 8, p = 0.67)

$Var(X_1 + X_3) = npq = 8 \bullet 0.67 \bullet 0.33 = 1.7688$

Example
Ezra maintains a server for his company's website. He knows that in the worldwide computer landscape, 65% of web surfers use operating system A, 12% use operating system B, 17% use operating system C, and 6% use operating system D. Ezra believes that clients visiting his website will come randomly from this population. In the next day, 20 clients visit his website.

(a.) Find the probability that precisely...
    14 clients use operating system A
    2 clients use operating system B
    4 clients use operating system C

0 clients use operating system D

(b.) What is the distribution for the number of clients (out of 20) using operating systems B, C, or D?

Solutions:
(a.)
We must find the probability that $\vec{X} = (14, 2, 4, 0)$.
We know the vector of probabilities: $\vec{p} = (0.65, 0.12, 0.17, 0.06)$.

$$f(x_1, x_2, x_3, x_4) = f(14, 2, 4, 0) = \binom{20}{14, 2, 4, 0} \cdot 0.65^{14} 0.12^2 0.17^4 0.06^0$$
$$= 581,400 \cdot 0.000000028903 = 0.016804$$

(b.)
(number of clients using systems B, C, or D) = $(X_2 + X_3 + X_4)$.
Pr(an outcome rests in category 2, 3, or 4) = $p_2 + p_3 + p_4 = 0.35$.
The number of trials is still 20.

$(X_2 + X_3 + X_4) \sim$ Binomial(n = 20, p = 0.35)

Chapter 7: Common Continuous Distributions

Several continuous distributions which appear quite often in probability have formal names. Each distribution has generic formulas for its probability density function, expected value, variance, and moment-generating function. The particular distributions encountered in this chapter include the continuous uniform distribution which we encountered earlier, the gamma and exponential distributions, and the beta distribution. Perhaps the most commonly used continuous distribution is the normal distribution.

Section 7-1: Continuous Uniform Distribution Revisited

Using tools developed in the last few chapters, we can elaborate on the continuous uniform distribution. Recall that a variable with this distribution has a pdf which is constant (horizontally level) over the interval [a, b]. All values within [a, b] are equally likely.

The expected value of a uniformly distributed random variable is the midpoint of the interval over which the pdf is positive. You can simply average the left and right endpoints to compute E(X).

Formulas for the Continuous Uniform Distribution

$$E(X) = \frac{a+b}{2}$$

$$Var(X) = \frac{(b-a)^2}{12}$$

$$\psi(t) = \frac{e^{bt} - e^{at}}{t(b-a)}$$

Example
Derive the variance formula for a continuous uniform distribution over the interval [a, b].

Solution:

$$E(X^2) = \int_a^b x^2 \cdot f(x)\, dx = \frac{1}{b-a} \cdot \int_a^b x^2\, dx = \frac{1}{b-a} \cdot \left[\frac{1}{3}x^3\right]_a^b = \frac{1}{b-a} \cdot \frac{1}{3} \cdot (b^3 - a^3)$$

$$Var(X) = E(X^2) - E(X)^2 = \frac{(b^3 - a^3)}{3(b-a)} - \frac{(a+b)^2}{4} = \frac{4(b^3 - a^3)}{12(b-a)} - \frac{3(a+b)^2(b-a)}{12(b-a)}$$

$$= \frac{4(b^3 - a^3) - 3(a^2 + 2ab + b^2)(b-a)}{12(b-a)}$$

$$= \frac{4(b^3 - a^3) - 3(a^2 b + 2ab^2 + b^3 - a^3 - 2a^2 b - ab^2)}{12(b-a)}$$

$$= \frac{4b^3 - 4a^3 - 3a^2 b - 6ab^2 - 3b^3 + 3a^3 + 6a^2 b + 3ab^2}{12(b-a)}$$

$$= \frac{b^3 - a^3 + 3a^2 b - 3ab^2}{12(b-a)} = \frac{(b-a)^3}{12(b-a)} = \frac{(b-a)^2}{12}$$

Example
TP Oil Industries opened a new well and believe that a uniform distribution will describe annual revenue from the well. Revenue should be somewhere between $120 and $190 million. What is the expected revenue and the standard deviation of revenue?

Solutions:
Let X = revenue per year.
X ~ Uniform(a = 120,000,000; b = 190,000,000)

$$E(X) = \frac{120{,}000{,}000 + 190{,}000{,}000}{2} = \$155{,}000{,}000$$

$$Var(X) = \frac{(190{,}000{,}000 - 120{,}000{,}000)^2}{12} = 4.083333 * 10^{14}$$

$$\sigma_X = \sqrt{Var(X)} = \$20{,}207{,}259$$

Example
Mitch is the district manager for a popular home improvement store chain. He oversees eight stores. The stores have a problem with forklift-related accidents. He would like to estimate the amount of money in lawsuits from the accidents over a year. He believes that the quantity of money in lawsuits for any given store during a year is uniformly distributed between \$500,000 and \$1,750,000. Assume that the loss experience for any given store is independent from that of other stores.

(a.) Find the expected value of lawsuits in the district over the next 3 years.
(b.) What is the variance and standard deviation in the monetary value of lawsuits over the next 2 years?

Solutions:
Let X_i = losses for i^{th} store
$X_i \sim$ Uniform(500000, 1750000)

Let T = total losses for the district = $X_1 + X_2 + \ldots + X_8$

(a.) $E(3T) = 3 * E(T) = 3 * 8 * E(X_i) = 24 * \dfrac{500{,}000 + 1{,}750{,}000}{2} = 27{,}000{,}000$

(b.) $Var(2T) = Var(2(X_1 + X_2 + \ldots + X_8)) = Var(2X_1 + 2X_2 + \ldots + 2X_8)$
$= 4Var(X_1) + 4Var(X_2) + \ldots + 4Var(X_8)$
$= 8 * 4Var(X_i) = 32 * Var(X_i) = 32 * \dfrac{(1{,}750{,}000 - 500{,}000)^2}{12}$
$= 4.166667 * 10^{12}$

stddev(2T) = 2,041,241

Section 7-2: Normal Distribution

The normal distribution is a continuous distribution. A variable that is normally distributed can assume any real value between $-\infty$ and $+\infty$. A normal distribution's pdf, when graphed, has the appearance of a bell curve. Just two parameters describe the distribution—the mean and standard deviation.

Parameters for the Normal Distribution
μ = population mean = $E(X)$
σ = population standard deviation = $\sqrt{Var(X)}$

The parameter σ must exceed 0, and μ can be any real number. Since the variance is easily derivable from the standard deviation, you could also use σ^2 as the second parameter. The pdf's are unimodal with just one hill. The mode of the normal distribution occurs at $x = \mu$, where the pdf reaches its global maximum. The normal distribution's pdf is symmetric about the vertical line $x = \mu$. The symmetric nature means that $f(\mu + x') = f(\mu - x')$. A normal distribution has just one median at $x = \mu$. Half the probability lies below μ and half the probability lies above μ.

For a normal distribution,
 expected value = median = mode

The pdf's differ in the sharpness of their peaks. When the variance is large, the pdf looks like a gradually sloping bump. On the other hand, a very small variance means that probability is concentrated close to the mean, and the pdf will have a very sharp peak and tails that seem to approach the x-axis quickly. The diagram below shows the pdf's for two normal distributions. Both distributions have the same mean, but a lower variance causes the pdf to spike.

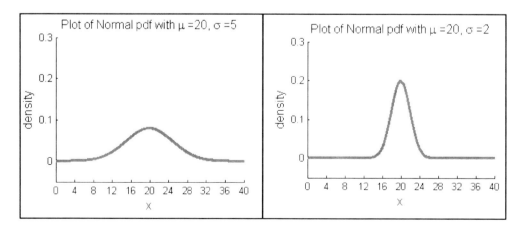

The pdf for a normal distribution is defined below. The density function is valid over all real x-values. Note the use of the z-value, which is a standard normal value (to be discussed later).

Probability Density Function for the Normal Distribution

$$f(x) = \frac{1}{\sqrt{2\pi} \cdot \sigma} \cdot e^{\wedge}\left[-\frac{1}{2}z^2\right]$$

where $z = \frac{x-\mu}{\sigma}$

Moment-Generating Function for the Normal Distribution

$$\psi(t) = e^{\wedge}\left[\mu t + \frac{1}{2}\sigma^2 t^2\right]$$

Probability X lies within k standard deviations of the mean:

k	$\Pr(\mu - k\sigma \leq X \leq \mu + k\sigma)$
1	0.6826
2	0.9546
3	0.9974
4	≈ 1

Standard Normal Distribution

The standard normal distribution is a specific type of normal distribution with the parameters fixed at $\mu = 0$ and $\sigma = 1$. The pdf for the standard normal distribution is a single bell curve with a peak over x = 0. The tails approach the x-axis quite rapidly.

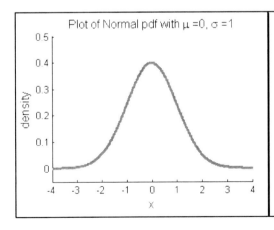
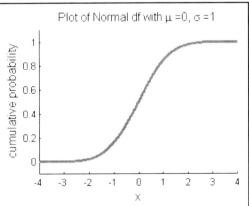

Probability Density Function for the Standard Normal Distribution

$$f(x) = \frac{1}{\sqrt{2\pi}} \cdot e^{\wedge}\left[-\frac{1}{2}x^2\right] = \phi(x)$$

A lowercase ϕ symbolizes the pdf of the standard normal distribution, while an uppercase Φ marks the df of the standard normal distribution.

The standard normal distribution is an essential tool to actually calculating probabilities from another normal distribution. To find the probability that X lies within some interval, you could integrate the general pdf formula for a Normal distribution, but it becomes tedious to do so. A simpler method to compute cumulative probabilities for any normal distribution is to convert x-values into standard normal z-values and then look-up cumulative probability values in a table. Converting x to z can be done with:

$$z = \frac{x - \mu}{\sigma}$$

The z-value reveals the number of standard deviations that x lies from its expected value. $Z \sim stdNormal(\mu = 0, \sigma = 1)$.

Example

Suppose X is distributed normally with mean 50 and variance 16. Find the following probabilities:

(a.) $Pr(X = 55)$ (b.) $Pr(X > 58)$ (c.) $Pr(45 < X < 52)$ (d.) $Pr(X < 48)$

Solutions:
(a.) X is a continuous random variable, so the probability that X takes a particular value must be 0.

(b.)
$$z = \frac{58 - 50}{4} = 2 \quad \rightarrow \quad Pr(X > 58) = Pr(Z > 2) = 1 - 0.9773 = 0.0227$$

(c.)
$$z_1 = \frac{45 - 50}{4} = -1.25 \quad z_2 = \frac{52 - 50}{4} = 0.5$$

$$Pr(45 < X < 52) = Pr(-1.25 < Z < 0.5) = Pr(Z < 0.5) - Pr(Z < -1.25)$$
$$= 0.6915 - (1 - 0.8944) = 0.5859$$

Digital Actuarial Resources — *Comprehensive Probability Review for Actuarial Exams*

(d.)
$$\Pr(X < 48) = \Pr\left(Z < \frac{48-50}{4}\right) = \Pr(Z < -0.5) = 1 - 0.6915 = 0.3085$$

Example

You are trying to fit a normal distribution to a collection of data. You have estimated the expected value of each sample point to be 75. You also believe that the chance that the random variable exceeds 90 is 30%. What is the variance for this distribution?

Solution:
$\Pr(X > 90) = 0.3 = \Pr(Z > z_1)$
What z-value satisfies this equation?

Using the quantile function which is implicitly provided in the standard normal table, we find that the chance Z exceeds 0.525 is about 30%. Therefore, let $z_1 = 0.525$.

$$0.525 = \frac{90-75}{\sigma} \quad \rightarrow \quad \sigma = 28.5714$$
$$Var(X) = 816.327$$

Lognormal Distribution

If a random variable X does not have a normal distribution, the natural logarithm of the variable might still be normally distributed. When $\ln(X) \sim Normal(\mu, \sigma^2)$, the distribution of X can be labeled as $X \sim LogNormal(\mu, \sigma^2)$.

Distribution Function for the Lognormal Distribution

$$F(x) = \Pr\left(Z < \frac{\ln(x) - \mu}{\sigma}\right)$$

Linear Combinations of Normal Distributions

In general, when you have a group of normally distributed random variables which are all mutually independent, the sum of the variables also has a normal distribution.

Sum of Many Normally Distributed Random Variables

Let $X_1 \sim Normal(\mu_1, \sigma_1)$, $X_2 \sim Normal(\mu_2, \sigma_2)$, ..., $X_n \sim Normal(\mu_n, \sigma_n)$.

All the X_i's are independent from each other.
Let $Y = X_1 + X_2 + ... + X_n$.

Y must have a normal distribution with
> mean = $\mu_1 + \mu_2 + ... + \mu_n$
> variance = $\sigma_1^2 + \sigma_2^2 + ... + \sigma_n^2$

Example
James, Tyrone, and De'Shawn are loading a car for a road trip. They are trying to determine the combined weight of their luggage. Each person packs independently of the others. The weight of each person's luggage follows a normal distribution:

$J \sim Normal(\mu_J = 80, \sigma_J = 7)$
$T \sim Normal(\mu_T = 35, \sigma_T = 3)$
$D \sim Normal(\mu_D = 126, \sigma_D = 15)$

(a.) What is the chance that their combined luggage weighs over 263 pounds?
(b.) Find the probability that their combined luggage weighs between 215 and 275 pounds?

Solutions:
Let $Y = J + T + D$.
Y~Normal(mean = 80 + 35 + 126, var = 49 + 9 + 225) = Normal(mean = 241, var = 283)

(a.)
$$\Pr(Y > 263) = \Pr\left(Z > \frac{263 - 241}{\sqrt{283}}\right) = \Pr(Z > 1.31) \approx 1 - 0.9049 = 0.0951$$

(b.)
$$z_1 = \frac{215 - 241}{\sqrt{283}} = -1.55 \qquad z_2 = \frac{275 - 241}{\sqrt{283}} = 2.02$$

$\Pr(215 < Y < 275) = \Pr(-1.55 < Z < 2.02) = \Pr(Z < 2.02) - \Pr(Z < -1.55)$
$\qquad = 0.9783 - 0.0606 = 0.9177$

General Linear Combination Formula for Many Normally Distributed Random Variables

Let $X_1 \sim Normal(\mu_1, \sigma_1)$, $X_2 \sim Normal(\mu_2, \sigma_2)$, ..., $X_n \sim Normal(\mu_n, \sigma_n)$.

All the X_i's are independent from each other.
Let $Y = a_1 X_1 + a_2 X_2 + ... + a_n X_n + b$.

Y must have a normal distribution with
$$\text{mean} = a_1 \mu_1 + a_2 \mu_2 + ... + a_n \mu_n + b$$
$$\text{variance} = a_1^2 \sigma_1^2 + a_2^2 \sigma_2^2 + ... + a_n^2 \sigma_n^2$$

Example

Coconuts, Bananas, and Alpha Male are three junior salesmen working for commission wages. The manager needs to predict the total wages for his salesmen over the next year. Each works on a different commission rate (a percentage of total sales). The amount of money in sales per year for any individual worker is normally distributed. The distribution for each worker is:

Alpha Male: mean = 85,000, var = 4,000,000, commission rate = 30%
Bananas: mean = 145,000, var = 12,250,000, commission rate = 37%
Coconuts: mean = 93,000, var = 2,890,000, commission rate = 28%

The manager must also pay a flat rate of $5,000 each year to simply keep the employees on the books. Assume the salesmen work independently of each other.

(a.) Find the distribution for the total cost of the employees per year.
(b.) What is the probability the total cost exceeds $112,000?

Solutions:
(a.)
Let $Y = a_A X_A + a_B X_B + a_C X_C + 5000$
$\downarrow$
$Y = 0.3 X_A + 0.37 X_B + 0.28 X_C + 5000$

Since X_A, X_B, and X_C are normally distributed, and Y is a linear combination of the X's, and the X's are independent, Y must also be normally distributed. The parameters for the distribution are:

mean = $0.3 * 85,000 + 0.37 * 145,000 + 0.28 * 93,000 + 5,000 = 110,190$

variance = $0.3^2 * 4,000,000 + 0.37^2 * 12,250,000 + 0.28^2 * 2,890,000 = 2,263,601$

(b.)

$$\Pr(Y > 112{,}000) = \Pr\left(Z > \frac{112{,}000 - 110{,}190}{\sqrt{2{,}263{,}601}}\right) = \Pr(Z > 1.20) = 1 - 0.8849 = 0.1151$$

Section 7-3: Correction for Continuity

The correction for continuity can be applied to any situation where a variable which is truly discrete is being modeled with a continuous distribution. This correction is typically utilized when the random variable can assume a set of integers without separation between the values. In many real-life problems, a continuous pdf will closely match a discrete variable's pf. Rather than specifying a large probability function (for perhaps thousands of x-values), the statistician can provide a compact, continuous function. Unfortunately, integrating the pdf to uncover the probability of a region is not the same as summing discrete probability values over the interval. One unavoidable source of error in a probability estimate is that the curve tends to overestimate probability in some regions and underestimate probability in others. Another type of error which is fixable occurs around the endpoints of an interval. A continuous approximation does not consider the fact that in the discrete case, the probability value at any particular x-value can be represented by a bar of width 1 over the x-value. By extending the interval a bit on both sides of an x-value, we can include all values which fit in that x-value's "bin." We need to be cautious when considering probability at endpoints of an interval.

An ideal example of committing the discrete-to-continuous error occurs when the normal distribution is used to approximate a distribution that is actually discrete. Let X be a discrete random variable which can take all integers with positive probability. A normal distribution with mean = 15 and var = 3 models X's distribution. The diagram below shows the actual bar plot (probability function values) and the approximated continuous pdf. Each bar is of width 1 and is centered over a particular x-value. The probability that X =c is found by computing the area of the bar above c. Since the bar-width is 1, the area of a bar equals its height. The pdf clearly overestimates and underestimates the true probability at certain x-values.

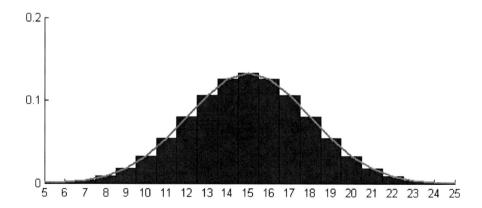

When finding the probability of a particular x-value using the continuous pdf, add and subtract 0.5 from that x-value to form the bounds of integration. Recall that for a continuous random variable, $\Pr(X = c) = 0$ for a constant c. We must expand the bar-width to 1 over $X = c$. The area of a bar corresponds to probability. If the bar width is a constant 1, then the height of the bar gives the actual probability value.

Probability of a Single Value Using the Correction for Continuity
Suppose X is a discrete random variable which has a distribution being described by a continuous pdf, f(x). The probability at $X = c$ is:

$$\Pr(X = c) = \int_{c-(1/2)}^{c+(1/2)} f(x)\, dx$$

When finding the probability over an interval of x-values, you need to make a slight adjustment to each endpoint to correct for continuity. Simply reduce the lower endpoint by 0.5 and increase the upper endpoint by 0.5.

Probability over an Interval Using the Correction for Continuity
Suppose X is a discrete random variable which has a distribution being described by a continuous pdf, f(x). The probability X lies within the interval [a, b] is:

$$\Pr(a \leq X \leq b) = \int_{a-(1/2)}^{b+(1/2)} f(x)\, dx$$

Deciding to not apply the correction for continuity usually results in an underestimate of probability in a region.

Probability over Unbounded Intervals Using the Correction for Continuity
Suppose X is a discrete random variable which has a distribution being described by a continuous pdf, f(x). Then,

$$\Pr(X < c) = \Pr(X \leq c - 1) = \int_{-\infty}^{c-(1/2)} f(x)\, dx$$

$$\Pr(X \leq c) = \int_{-\infty}^{c+(1/2)} f(x)\, dx$$

Example
The number of people registering for a computer vision conference is normally distributed with mean 340 and standard deviation 19. Find the chance that between 313 and 351 people, inclusive, actually register for the conference using...

(a.) no correction for continuity
(b.) the continuity correction

Solutions:
The number of people coming to an event is a discrete variable, since fractions of people are meaningless.

(a.) Apply the simple definition of the normal distribution, wrongly assuming that the random variable is continuous.

Let X = number of people registered for conference.
X ~ Normal(μ =340, σ =19).

$$\Pr(313 < X < 351) = \Pr\left(\frac{313-340}{19} < Z < \frac{351-340}{19}\right) = \Pr(-1.42 < Z < 0.58)$$
$$= \Pr(Z < 0.58) - \Pr(Z < -1.42) = 0.719 - 0.0778 = 0.6412$$

(b.)
Subtract ½ from the lower endpoint and add ½ to the upper endpoint.

$$\Pr(312.5 < X < 351.5) = \Pr\left(\frac{312.5-340}{19} < Z < \frac{351.5-340}{19}\right) = \Pr(-1.45 < Z < 0.61)$$
$$= \Pr(Z < 0.61) - \Pr(Z < -1.45) = 0.7291 - 0.0735 = 0.6556$$

The estimate of the probability using the correction for continuity is slightly higher than the other probability estimate ignoring the correction.

Example
The number of bull riders entering a rodeo is approximately normally distributed with mean 22 and standard deviation 5. Compute the probability that between 20 and 25 riders, inclusive, enter the competition by using...

(a.) no correction for continuity
(b.) the continuity correction

Solutions:
Let X = # cowboys entering the contest

(a.) $\Pr(20 \leq X \leq 25) = \Pr\left(\dfrac{20-22}{5} \leq Z \leq \dfrac{25-22}{5}\right) = \Pr(-0.4 \leq Z \leq 0.6)$

$\qquad = 0.7257 - 0.3446 = 0.3811$

(b.) $\Pr(20 \leq X \leq 25) \quad \rightarrow \quad \Pr(19.5 \leq X \leq 25.5)$

$\qquad = \Pr\left(\dfrac{19.5-22}{5} \leq Z \leq \dfrac{25.5-22}{5}\right) = \Pr(-0.5 \leq Z \leq 0.7)$

$\qquad = 0.7580 - 0.3085 = 0.4495$

Section 7-4: Central Limit Theorem

The central limit theorem claims that certain statistics of a random sample will follow a normal distribution. In particular, the sum and mean of a random sample will be roughly normally distributed. The sample mean is proportional to the sample sum with factor (1 / n). In order for the central limit theorem to apply, the sample size needs to be large. A "large" sample typically has a sample size equal to or greater than 30.

Please note that the X_i's must all be independent and identically distributed. That is, all the X_i's must follow the same distribution. The X_i's could come from any distribution. They do not need to be normally distributed individually. For example, all the Xi's could be uniformly distribution on [0, 1], or they could follow the same Poisson distribution.

The sum of many iid random variables has a distribution that is roughly normal.

Central Limit Theorem for a Sum of Independent, Identically Distributed Random Variables

Let $X_1, X_2, \ldots, X_n$ be iid sample points. Each X_i has mean μ and variance σ^2.
The sum of the variables, $X_1 + X_2 + \ldots + X_n$, has a distribution that is approximately normal with

$\qquad$ mean = $n\mu$ $\quad$ and $\quad$ variance = $n\sigma^2$

Example
Uriah collects a random sample of size 35. Each sample point comes from the continuous uniform distribution on the interval [12, 18].
(a.) What is the distribution for the sum of the sample?
(b.) What is the chance the sum of the sample surpasses 540?

Solutions:
(a.)
Let $Y = X_1 + X_2 + \ldots + X_{35}$

$$E(X_i) = \mu = \frac{12+18}{2} = 15 \qquad Var(X_i) = \sigma^2 = \frac{(18-12)^2}{12} = 3$$

Y has a distribution that is approximately normal with…
$E(Y) = 35*15 = 525$
$Var(Y) = 35*3 = 105$

(b.)
$$\Pr(Y > 540) = \Pr\left(Z > \frac{540-525}{\sqrt{105}}\right) = \Pr(Z > 1.46) = 0.0721$$

Example
A random variable X has this pdf:

$$f(x) = \begin{cases} 30x^4(1-x), & \text{if } 0 < x < 1 \\ 0, & \text{else} \end{cases}$$

The experimenter collects 40 independent, separate observations of X.
Let Y denote the sum of the random sample.

(a.) What is the distribution of Y?
(b.) What is the probability that Y is between 29 and 29.5?

Solutions:
(a.) First, let's find the expected value and variance of a X_i observation.

$$E(X) = 30\int_0^1 x^5 - x^6 \, dx = 30 * \left[\frac{1}{6}x^6 - \frac{1}{7}x^7\right]_0^1 = 30 * \left(\frac{1}{6} - \frac{1}{7}\right) = \frac{5}{7}$$

$$E(X^2) = 30\int_0^1 x^6 - x^7 \, dx = 30 * \left[\frac{1}{7}x^7 - \frac{1}{8}x^8\right]_0^1 = 30 * \left(\frac{1}{7} - \frac{1}{8}\right) = \frac{15}{28}$$

$$Var(X) = \frac{15}{28} - \left(\frac{5}{7}\right)^2 = \frac{5}{196}$$

Y has approximately a normal distribution with parameters…

$$E(Y) = 40*(5/7) = 200/7$$
$$Var(Y) = 40*(5/196) = 50/49$$

(b.)
$$Pr(29 < Y < 29.5) = Pr\left(\frac{29-(200/7)}{\sqrt{50/49}} < Z < \frac{29.5-(200/7)}{\sqrt{50/49}}\right)$$
$$= Pr(0.42 < Z < 0.92) = 0.8212 - 0.6628 = 0.1584$$

The central limit theorem also states that as the sample size becomes sufficiently large, the sample mean will approach the population mean. In a previous section, we found that the mathematical average of a random sample has a distribution which is nearly normal.

Central Limit Theorem for the Sample Mean

Let $X_1, X_2, \ldots, X_n$ be iid sample points. Each X_i has mean μ and variance σ^2.
Let $\overline{X}_n$ be the sample mean of the observations.
Then, $\overline{X}_n$ has a distribution that is approximately normal with

$$E(\overline{X}_n) = \mu \qquad Var(\overline{X}_n) = \frac{\sigma^2}{n}$$

The sample mean being normally distributed intuitively makes sense. Many samples from the same distribution should have a mathematical average close to the actual expected value from the original distribution. Sometimes the sample mean will be slightly higher than the true population mean. With equal probability, the sample mean could be slightly lower than the population mean. However, we would not anticipate a sample mean extremely far from the population mean. In the end, the bell curve for a normal pdf adequately describes the behavior of the sample mean.

The central limit theorem applied to the sample mean can be formally stated in a theorem from Lindeberg and Levy. You can evaluate probabilities that $\overline{X}_n$ lies within the range [a, b] by converting a and b to standard normal values.

Lindeberg and Levy's Central Limit Theorem

Let $\overline{X}_n$ be the sample mean of n iid sample points.

$$\lim_{n \to \infty} Pr(a \le \overline{X}_n \le b) = Pr\left(\frac{a-\mu}{\sqrt{\sigma^2/n}} \le Z \le \frac{b-\mu}{\sqrt{\sigma^2/n}}\right) = \Phi\left(\frac{b-\mu}{\sqrt{\sigma^2/n}}\right) - \Phi\left(\frac{a-\mu}{\sqrt{\sigma^2/n}}\right)$$

where Z has a standard normal distribution.

Example
TP Oil Industries operates 80 wells in a region. The amount of revenue that each well brings annually has a continuous uniform distribution with minimum = $58 million and maximum = $124 million. The executives are concerned about the average amount of revenue that a well brings in.

(a.) Find the distribution of the sample mean.
(b.) What is the chance that the average amount of revenue is between $90 million and $91.5 million?

Solutions:
(a.) $\overline{X}_n$ ~ Normal with the following parameters:

$$\text{mean} = \frac{58{,}000{,}000 + 124{,}000{,}000}{2} = \$91{,}000{,}000$$

$$\text{var} = \sigma^2/n = \frac{(124{,}000{,}000 - 58{,}000{,}000)^2}{12} \Big/ 80 = 4.5375 * 10^{12}$$

$$\text{stddev}(\overline{X}_n) = \$2{,}130{,}141$$

(b.)
$\Pr(90{,}000{,}000 < \overline{X}_n < 91{,}500{,}000) =$

$$= \Pr\left(\frac{90{,}000{,}000 - 91{,}000{,}000}{2{,}130{,}141} < Z < \frac{91{,}500{,}000 - 91{,}000{,}000}{2{,}130{,}141}\right)$$

$= \Pr(-0.47 < Z < 0.23)$
$= \Pr(Z < 0.23) - \Pr(Z < -0.47) = 0.591 - 0.3192 = 0.2718$

Example
Each X_i in a random sample has this discrete probability function:

x	f(x)
4	0.28
8	0.05
12	0.46
16	0.21

The sample size is 24.

What is the chance that $\overline{X}_n$ is between 10 and 11, inclusive, without using the correction for continuity?

Solution:
First, find the expected value and variance for an individual trial, X_i.

$E(X_i) = 4*0.28 + 8*0.05 + 12*0.46 + 16*0.21 = 10.4$

$E(X_i^2) = 4^2*0.28 + 8^2*0.05 + 12^2*0.46 + 16^2*0.21 = 127.68$

$Var(X_i) = E(X_i^2) - E(X_i)^2 = 127.68 - 10.4^2 = 19.52$

$\overline{X}_n \sim$ Normal with the following parameters:

mean = $E(X_i) = 10.4$

var = $Var(X_i)/n = 19.52/24 = 0.81333$
stddev = 0.90185

$Pr(10 \leq \overline{X}_n \leq 11) = Pr\left(\dfrac{10-10.4}{0.90185} < Z < \dfrac{11-10.4}{0.90185}\right) = Pr(-0.44 < Z < 0.67)$

$= 0.4186$

Example

Each X_i in a random sample has this discrete probability function:

x	f(x)
1	0.3
2	0.2
3	0.2
4	0.3

The sample size is 18. Let Y be the sum of all trials.

(a.) What is the chance that Y is between 42 and 50, inclusive, without employing the correction for continuity?

(b.) Find the probability from part a by employing the correction for continuity.

Solutions:
First, gather statistics on an individual X_i trial.

$E(X_i) = 2.5$, $E(X_i^2) = 7.7$, $Var(X_i) = 1.45$

Y ~ Normal with parameters...
  mean = n * E(X_i) = 18 * 2.5 = 45
  var = n * Var(X_i) = 18 * 1.45 = 26.1

(a.)
$$\Pr(42 \leq Y \leq 50) = \Pr\left(\frac{42-45}{\sqrt{26.1}} \leq Z \leq \frac{50-45}{\sqrt{26.1}}\right) = \Pr(-0.59 \leq Z \leq 0.98) = 0.5589$$

(b.)
$\Pr(42 \leq Y \leq 50) \rightarrow \Pr(41.5 \leq Y \leq 50.5)$
$$= \Pr\left(\frac{41.5-45}{\sqrt{26.1}} \leq Z \leq \frac{50.5-45}{\sqrt{26.1}}\right) = \Pr(-0.69 \leq Z \leq 1.08) = 0.6148$$

Example
The random variable X is continuous with this pdf:

$$f(x) = \begin{cases} 9x^8, & \text{if } 0 < x < 1 \\ 0, & \text{else} \end{cases}$$

The experimenter collects 60 independent sample points.
(a.) What is the distribution of the sample mean?
(b.) Compute the chance that the sample mean is above 0.89.

Solutions:
(a.)
First, find the expected value and variable of a single observation.

$$E(X) = \int_0^1 9x^9 \, dx = 9\left[\frac{1}{10}x^{10}\right]_0^1 = 9/10$$

$$E(X^2) = \int_0^1 9x^{10} \, dx = 9\left[\frac{1}{11}x^{11}\right]_0^1 = 9/11$$

$Var(X) = 9/11 - (9/10)^2 = 9/1100$

The sample mean has roughly a normal distribution with
  $E(\overline{X}_n) = \mu = 9/10$

$$Var(\overline{X}_n) = \frac{\sigma^2}{n} = \frac{9/1100}{60} = 0.00013636$$

(b.)

$$\Pr(\overline{X}_n > 0.89) = \Pr\left(Z > \frac{0.89 - 0.9}{\sqrt{0.00013636}}\right) = \Pr(Z > -0.86) = 0.8051$$

Section 7-5: Bivariate Normal Distribution

A bivariate normal distribution involves two variables, each of which has a marginal distribution that is normal. The two variables could be dependent or independent with each other. If you can show that X_1 and X_2 each have an individual normal distribution, and $Cov(X_1, X_2)$ is defined, then X_1 and X_2 automatically have a bivariate normal distribution. The joint pdf for this distribution is a bit complicated:

Joint pdf for Bivariate Normal Random Variables

$$f(x_1, x_2) = \frac{1}{2\pi\sqrt{(1-\rho^2)}\sigma_1\sigma_2} * e^{\wedge}\left[-\frac{1}{2(1-\rho^2)}\left(z_1^2 - 2\rho z_1 z_2 + z_2^2\right)\right]$$

where $z_1 = \frac{x_1 - \mu_1}{\sigma_1}$ and $z_2 = \frac{x_2 - \mu_2}{\sigma_2}$

and $x_1, x_2 \in R$

The bivariate normal distribution has a special rule for the relationship between the independence of its two variables and their correlation. In the typical case of any two random variables, when the variables are independent, then $\rho = 0$. The converse of the statement is typically untrue—if $\rho = 0$ for two random variables, then they could still be dependent upon each other (perhaps dependent in a nonlinear fashion). Surprisingly, in the case of two random variables which share a bivariate normal distribution, if $\rho = 0$, then the variables are independent.

Independence Condition for Variables in a Bivariate Normal Distribution
Let X_1 and X_2 belong to a bivariate normal distribution. Then,

$\rho = 0$ iff X_1 and X_2 are independent

As further proof that $\rho = 0$ implies independence, let's show that when $\rho = 0$, the joint pdf can be factored into $f_1(x_1) * f_2(x_2)$:

$$f(x_1, x_2) = \frac{1}{2\pi\sqrt{(1-0^2)}\sigma_1\sigma_2} * e^{\wedge}\left[-\frac{1}{2(1-0^2)}\left(z_1^2 - 2*0*z_1z_2 + z_2^2\right)\right]$$

$\downarrow$

$$f(x_1, x_2) = \frac{1}{2\pi\sigma_1\sigma_2} * e^{\wedge}\left[-\frac{1}{2}\left(z_1^2 + z_2^2\right)\right]$$

$\downarrow$

$$f(x_1, x_2) = \frac{1}{\sqrt{2\pi}\sigma_1} * e^{\wedge}\left[-\frac{1}{2}z_1^2\right] * \frac{1}{\sqrt{2\pi}\sigma_2} * e^{\wedge}\left[-\frac{1}{2}z_2^2\right] = f_1(x_1) * f_2(x_2)$$

Conditional Probability Distributions

We will often want to predict one variable after being given the value of the second variable from a bivariate normal distribution. The conditional random variable, such as ($X_1 \mid X_2 = x_2$), is also normally distributed. The parameters for the conditional random variable depend on the parameters from X_1 and X_2, along with the observed variable's value. The value of ρ will reveal how strongly the two variables vary together. A high correlation leads to more accurate predictions of one variable when given the value of the other variable. Likewise, the variance in the prediction will be lower when the correlation coefficient is higher.

Conditional pdf of ($X_2 \mid X_1 = x_1$)
Suppose X_1 and X_2 share a bivariate normal distribution.
Suppose also that the mean and variance of each random variable is defined, and the correlation coefficient between the variables is defined.
Suppose also that X_1's value is already given, and you wish to find the distribution of X_2 conditional on X_1's value.
Then,

The random variable ($X_2 \mid X_1 = x_1$) has a normal distribution with parameters
mean = $\mu_2 + \rho\sigma_2 z_1$
variance = $(1 - \rho^2)\sigma_2^2$

Conditional pdf of ($X_1 \mid X_2 = x_2$)
Suppose X_1 and X_2 share a bivariate normal distribution.
Suppose also that the mean and variance of each random variable is defined, and the correlation coefficient between the variables is defined.

Suppose also that X_2's value is already given, and you wish to find the distribution of X_1 conditional on X_2's value.
Then,

The random variable $(X_1 \mid X_2 = x_2)$ has a normal distribution with parameters
mean = $\mu_1 + \rho \sigma_1 z_2$
variance = $(1 - \rho^2)\sigma_1^2$

Example
Itidal is a sociologist studying the relationship between how many hours students listen to music each week and their starting salaries in the work force. She has found that the two variables have a negative correlation such that increased music tends to occur with a lowered salary for any given student. In fact, the correlation coefficient between the two traits is -0.63. In the student population at a college, Itidal found that music use in a student is normally distributed with mean = 35 hours and variance = 12 hours. Starting salaries are also normally distributed with mean = $18,000 and standard deviation = $7,000.

(a.) Suppose Itidal examines student A. She does not know how much music student A listens to each week. What value should Itidal use for student A's expected starting salary?

(b.) Itidal is still examining student A and now finds that the student listens to 39 hours of music per week. What value should Itidal now use for the student's expected starting salary?

(c.) What is the MSE of the predicted salary in part a? What is the MSE of the predicted salary in part b? Did the extra information of knowing the student's music habits improve the prediction of salary?

Solutions:
Let X_1 = music use per week in hours
Let X_2 = starting salary in dollars
We don't need to build any pdf's for this example.

(a.)
$E(X_2) = \$18,000$

(b.)
$E(X_2 \mid X_1 = 39) = 18{,}000 - 0.63 * 7000 * \left(\dfrac{39 - 35}{\sqrt{12}} \right) = \$12{,}907.77$

(c.)

MSE of prediction from a = $7{,}000^2 = 49{,}000{,}000$

MSE of prediction from b = $(1-(-0.63)^2)*7{,}000^2 = 29{,}551{,}900$

Yes, knowing the amount of music moderately improved the prediction of salary. The MSE of the prediction fell by 19,448,100.

Example
Helmut is examining the connection between IQ test scores and number of books read so far in life for graduating seniors in a high school. He believes that an IQ score follows a normal distribution with mean = 100 and standard deviation = 15. In addition, the quantity of books read seems to obey a normal distribution with mean = 1,600 and standard deviation = 230. Helmut believes the correlation coefficient between the two variables is 0.87. Helmut is studying the abilities of a particular student, Dieter.

(a.) Prior to Dieter taking an IQ test, what is the expected number of books which Dieter has read?
(b.) Dieter scored 145 on the IQ test. What value should Helmut now use for Dieter's expected number of books read?
(c.) What is the improvement in the MSE of the predicted number of books Dieter read after we learn Dieter's IQ score?

Solutions:
Let X_1 = IQ score, $X_1 \sim$ Normal, $E(X_1) = 100$, stddev(X_1) = 15
Let X_2 = # books read, $X_2 \sim$ Normal, $E(X_2) = 1600$, stddev(X_2) = 230
$\rho = 0.87$

(a.)
$E(X_2) = 1600$

(b.)
$E(X_2 | X_1 = 145) = 1600 + 0.87 * 230 * \left(\dfrac{145-100}{15}\right) = 2200$

(c.)
MSE of prediction from part a = $230^2 = 52{,}900$

MSE of prediction from part b = $(1-0.87^2)230^2 = 12{,}859.99$

Reduction in MSE = 40,040.01

When X_1 and X_2 have a bivariate normal distribution, any linear combination of the two variables will also be normally distributed. In formal terms,

Digital Actuarial Resources   *Comprehensive Probability Review for Actuarial Exams*

Linear Combination of Variables in a Bivariate Normal Distribution

Suppose X_1 and X_2 have a bivariate normal distribution.
Let $Y = w_1 X_1 + w_2 X_2 + c$.
Then, Y has a normal distribution with...
  mean = $w_1 * E(X_1) + w_2 * E(X_2) + c$
  variance = $w_1^2 * Var(X_1) + w_2^2 * Var(X_2) + 2 w_1 w_2 \rho * stddev(X_1) * stddev(X_2)$

In particular, the sum or difference of X_1 and X_2 will follow a normal distribution.

Example

Heinrich, a professor of advanced medicine, is examining the relationship between a mother's weight and her daughter's weight. Heinrich has controlled for age differences between the mother and daughter. He has noticed a moderately strong correlation of 0.72 between the mother's weight and the daughter's weight. For any mother in the population, her weight is normally distributed with mean = 280 pounds and standard deviation = 60 pounds. Among the young daughters in the population, any daughter has a weight that is normally distributed with mean = 265 pounds and standard deviation = 52 pounds. The mother's weight and daughter's weight form a bivariate normal distribution.

What is the chance the mother weighs more than the daughter for any given selection?

Solution:
Let X_1 = mother's weight, X_2 = daughter's weight
Let $Y = X_1 - X_2$
We must find $Pr(X_1 > X_2) = Pr(X_1 - X_2 > 0) = Pr(Y > 0)$

Y must have a normal distribution with these parameters:
  mean = $E(X_1) - E(X_2) = 15$
  variance = $60^2 + 52^2 + 2(1)(-1) * 0.72 * 60 * 52 = 1{,}811.2$

$$Pr(Y > 0) = Pr\left(Z > \frac{0 - 15}{\sqrt{1{,}811.2}} \right) = Pr(Z > -0.35) = 0.6368$$

Example

A researcher believes that the variables X_1 and X_2 have a bivariate normal distribution. He has collected the following figures:

$E(X_1) = 130$, $Var(X_1) = 198$
$E(X_2) = 85$, $Var(X_2) = 57$
$\rho = -0.29$

Find the distribution for Y, where $Y = 6X_1 + 4X_2 - 40$.

Solution:
Y will have a normal distribution with the following parameters:

mean $= 6*130 + 4*85 - 40 = 1,080$
variance $= 6^2 *198 + 4^2 *57 + 2*6*4*(-0.29)*\sqrt{198}\sqrt{57} = 6,561.2006$

Section 7-6: Gamma Distribution

The gamma distribution is a continuous distribution which utilizes the gamma function. The distribution has two parameters, α and β, which both need to be greater than 0. The gamma distribution is useful when the random variable can only take values above 0.

Gamma Function
$$\Gamma(\alpha) = \int_0^\infty x^{\alpha-1} e^{-x} \, dx$$

where $\alpha > 0$

Evaluating the gamma function for most floating point values of alpha can be quite difficult. You would need to use an iterative numerical analysis technique to evaluate gamma under most non-integer parameters.

For $\alpha > 1$,
$\Gamma(\alpha) = (\alpha - 1) \bullet \Gamma(\alpha - 1)$

If α is a positive integer, then
$\Gamma(\alpha) = (\alpha - 1)!$

Important Values of the Gamma Function
$\Gamma(1/10) = 9.5135$
$\Gamma(2/10) = 4.5908$
$\Gamma(3/10) = 2.9916$
$\Gamma(4/10) = 2.2182$
$\Gamma(1/2) = \sqrt{\pi} = 1.7725$
$\Gamma(6/10) = 1.4892$

$\Gamma(7/10) = 1.2981$
$\Gamma(8/10) = 1.1642$
$\Gamma(9/10) = 1.0686$

To evaluate gamma with a floating point parameter above 1, you should reduce the expression using $\Gamma(\alpha) = (\alpha-1) \bullet \Gamma(\alpha-1)$ until the argument inside the remaining gamma function is between 0 and 1. You can then use the above table to compute the remaining gamma function.

Example
Evaluate the following:
(a.) $\Gamma(7)$ (b.) $\Gamma(3/2)$ (c.) $\Gamma(1)$ (d.) $\Gamma(5/2)$ (e.) $\Gamma(3.8)$ (f.) $\Gamma(4.3)$

Solutions:
(a.) $\Gamma(7) = (7-1)! = 720$

(b.) $\Gamma(3/2) = (1/2) * \Gamma(1/2) = \dfrac{\sqrt{\pi}}{2}$

(c.) $\Gamma(1) = 1$

(d.) $\Gamma(5/2) = (3/2) * \Gamma(3/2) = \dfrac{3}{2} * \dfrac{\sqrt{\pi}}{2} = \dfrac{3\sqrt{\pi}}{4}$

(e.)
$\Gamma(3.8) = 2.8 * \Gamma(2.8) = 2.8 * 1.8 * \Gamma(1.8) = 2.8 * 1.8 * 0.8 * \Gamma(0.8) =$
$= 4.032 * \Gamma(0.8) = 4.6941$

(f.)
$\Gamma(4.3) = 3.3 * \Gamma(3.3) = 3.3 * 2.3 * \Gamma(2.3) = 3.3 * 2.3 * 1.3 * \Gamma(1.3) =$
$= 3.3 * 2.3 * 1.3 * 0.3 * \Gamma(0.3) = 8.8554$

The gamma distribution's pdf consists of a few simple terms. The pdf is positive for x-values exceeding zero. The leading term of $\beta^{\alpha} / \Gamma(\alpha)$ evaluates to a real value. This term is particularly easy to calculate when α and β are integers. You should not be intimidated by $\Gamma(\alpha)$, especially if α is an integer. The pdf also contains a simple algebraic term involving x, along with an exponential term.

Probability Density Function for the Gamma Distribution

$$f(x) = \begin{cases} \dfrac{\beta^{\alpha}}{\Gamma(\alpha)} x^{\alpha-1} e^{-\beta x}, & \text{if } x > 0 \\ 0, & \text{else} \end{cases}$$

Several plots of the gamma pdf for various parameter values are shown below. The pdf resembles a skewed bell curve.

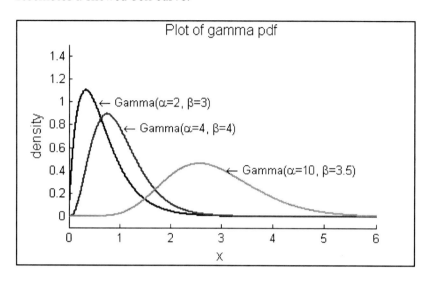

The short formulas below give the expected value, variance, and mgf for any gamma distribution.

Formulas for the Gamma Distribution

$$E(X) = \frac{\alpha}{\beta}$$

$$Var(X) = \frac{\alpha}{\beta^2}$$

$$\psi(t) = \left(\frac{\beta}{\beta - t}\right)^\alpha$$

Sum of Many Gamma Distributed Random Variables
Let $X_1, X_2, \ldots, X_n$ be gamma distributed random variables.
Each variable is independent from the others.
Suppose X_i has parameters α_i and β (all the variables have the same beta).
Let $Y = X_1 + X_2 + \ldots + X_n$.

Then, Y is gamma distributed with…
 alpha = $\alpha_1 + \alpha_2 + \ldots + \alpha_n$
 beta = β

Exponential Distribution

The exponential distribution is a special case of the gamma distribution with parameter $\alpha = 1$. The exponential distribution has just one parameter, β, which needs to exceed 0. In a Poisson process, the gaps in time between happenings are exponentially distributed.

The pdf for a generic exponential distribution is formed by plugging $\alpha = 1$ into the formula for a gamma distribution's pdf. The resulting pdf has just a coefficient and exponential term.

Probability Density Function for the Exponential Distribution
$$f(x) = \begin{cases} \beta e^{-\beta x}, & \text{if } x > 0 \\ 0, & \text{else} \end{cases}$$

A few plots of the exponential pdf are illustrated here. They reach their global maximums at ε, where ε is just slightly larger than 0.

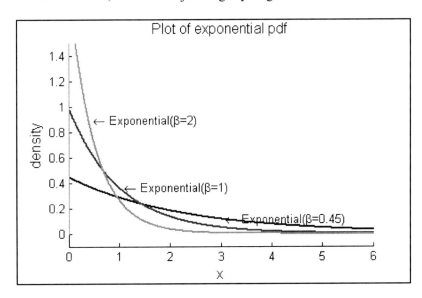

Distribution Function for the Exponential Distribution
$\Pr(X \leq x) = F(x) = 1 - e^{-\beta x}$ if $x > 0$

Example
What is the lower quartile, median, and upper quartile for a random variable named X which is exponentially distributed with $\beta = 4.5$?

Solutions:
A few methods exist to find the quartiles. You could integrate the pdf from 0 up to the unknown quartile, set the integral to the desired percentile (such as 0.25), and then solve for the unknown quartile. Another method is to use the distribution function. Yet another technique would be to solve for the quantile function and use this function directly.

Finding the lower quartile:

$$\int_0^a 4.5 e^{-4.5x} \, dx = 0.25 \quad \rightarrow \quad 4.5 * \left[-\frac{1}{4.5} e^{-4.5x} \right]_0^a = 0.25$$

$$-\left(e^{-4.5a} - e^0 \right) = 0.25 \quad \rightarrow \quad -e^{-4.5a} + 1 = 0.25$$

$$-e^{-4.5a} = -0.75 \quad \rightarrow \quad e^{-4.5a} = 0.75 \quad \rightarrow \quad -4.5a = \ln(0.75)$$

a = lower quartile = 0.06393

OR

$F(a) = 1 - e^{-4.5a} = 0.25$
$-e^{-4.5a} = -0.75$
$e^{-4.5a} = 0.75$
$-4.5a = \ln(0.75)$
a = 0.06393

OR

$F(x) = 1 - e^{-\beta x} = p$
$-e^{-\beta x} = p - 1$
$e^{-\beta x} = 1 - p$
$-\beta x = \ln(1 - p)$
$x = -\dfrac{\ln(1-p)}{\beta} = F^{-1}(p)$ = quantile function for any exponential distribution

$a = -\dfrac{\ln(0.75)}{4.5} = 0.06393$

Finding the median:
$$b = -\frac{\ln(0.5)}{4.5} = 0.15403$$

Finding the upper quartile:
$$c = -\frac{\ln(0.25)}{4.5} = 0.30807$$

Formulas for the Exponential Distribution

$$E(X) = \frac{1}{\beta}$$

$$Var(X) = \frac{1}{\beta^2}$$

$$\psi(t) = \frac{\beta}{\beta - t}$$

The exponential distribution is memoryless. In fact, it is the only continuous distribution with the memoryless property. The memoryless trait means that past history has no impact on the future likelihood of an event. The whole process essentially restarts at every moment.

The memoryless feature does not always describe all situations well. People frequently use the exponential distribution to model the lifetimes of products. An example of a random variable with an exponential distribution could be the lifetime of a computer chip. Clearly, the lifetime must be strictly positive. It is possible the chip never breaks down and the random variable approaches positive infinity. The computer chip endures wear-and-tear over time, degrading with every moment of use. However, the memoryless trait says that the chip restarts life every moment. All the past wear-and-tear can be disregarded. If the chip initially had an expected lifetime of 5 years, it will still have an expected use of 5 years from any moment in the future. The memoryless process does not consider the fact that the chip's future lifetime will be shortened from previous use.

The memoryless trait is evident when examining the times between occurrences of some phenomenon. Consider a pool of n objects, each of which must experience the phenomenon at some moment of time in the future. Let X_i be the time of occurrence for the ith object, and allow X_i to have an exponential distribution with parameter β. Assume that all the X_i's are iid so that they form a random sample. In addition, suppose that after we obtain all the values for the X_i's, we order them from smallest to largest. The gaps of time between the ordered occurrences will follow exponential distributions. Let Y_1 be the span of time from the start to the first occurrence, let Y_2 equal the gap of time from the first to second occurrence, let Y_3 be the amount of time from the second to third occurrence, and so on. Each Y_i has an exponential distribution with a unique parameter. At the start of the experiment, n objects exist, each with parameter β. As a result, Y_1 has an exponential distribution with parameter equal to $n\beta$. The objects basically restart life at every occurrence, so their parameters remain unchanged. The $(n-1)$ objects which did not die in the first round will return for the second round. Y_2 is exponentially distributed with parameter set at $(n-1)\beta$. The process will continue until the very last object remains. Even though this final object has been alive and active since the experiment's start, it will still restart and have parameter β for the time until its death. All the Y_i's are independent from each other.

Differences between Successive Orderings of Exponentially Distributed Variables in a Random Sample

Let $X_1, X_2, \ldots, X_n$ each have an exponential distribution with common parameter β. That is, the X_i variables come from a random sample.

Let Y_i = difference between $(i-1)^{th}$ and i^{th} sample points from among the X_i's, assuming the samples points were ordered from smallest to largest.

In particular,
Y_1 = minimum from among all the X_i's.
Y_2 = distance between first and second occurrences.
Y_3 = distance between second and third occurrences.

Then, Y_i is exponentially distributed with
beta = $(n-i+1)*\beta$

In particular,
$Y_1 \sim \text{Exp}(n\beta)$
$Y_2 \sim \text{Exp}((n-1)\beta)$
$Y_3 \sim \text{Exp}((n-2)\beta)$

Example

A new retail store hires 30 employees. Each employee will eventually leave the company for some reason (death due to workplace accident, natural death, seeking higher-level employment, etc.). Suppose each employee's time of service (in years) follows an exponential distribution with parameter $\beta = 0.2$. Assume the company will not hire replacements.

(a.) What is the chance that someone leaves the company within the first 6 months?
(b.) Two employees leave at the same moment after a week of work. What's the chance the third employee to quit does so in the next 4 months?

Solutions:

(a.)
Let Y_1 = time of first departure for an employee
$Y_1 \sim \text{Exp}(30 * 0.2 = 6)$

$\Pr(Y_1 < 0.5) = 1 - e^{-6*0.5} = 0.9502$

(b.)
Let Y_3 = time between second quitting and third quitting
Again, the process "restarts" so that the future time any employee will work is exponentially distributed with $\beta = 0.2$. However, we now have just 28 employees to deal with.

$Y_3 \sim \text{Exp}(28*0.2 = 5.6)$

$\Pr(Y_3 < 0.3333) = 1 - e^{-5.6*0.3333} = 0.8454$

Example
TP Oil Industries has four wells in a region. The company can really only handle one broken well at a time, and the firm needs 5 months to close any well that breaks. Once a well is broken, it cannot break again. Assume all the wells operate independently. The time until any well breaks is exponentially distributed with an expected time of 26 months. What's the chance that no two successive breaks happen within 5 months of each other?

Solution:
$\beta = 1/26$

We need to find this probability:
$\Pr(Y_2 > 5 \cap Y_3 > 5 \cap Y_4 > 5) = \Pr(Y_2 > 5) * \Pr(Y_3 > 5) * \Pr(Y_4 > 5)$

You can multiply the probability terms because the Y_i's are independent.

$Y_2 \sim \text{Exp}(3*(1/26) = 3/26))$
$Y_3 \sim \text{Exp}(2*(1/26) = 2/26))$
$Y_4 \sim \text{Exp}(1*(1/26) = 1/26))$

$\Pr(Y_2 > 5 \cap Y_3 > 5 \cap Y_4 > 5) = e^{-(3/26)*5} * e^{-(2/26)*5} * e^{-(1/26)*5}$
$= e^{-(15/26)} * e^{-(10/26)} * e^{-(5/26)}$
$= e^{-(30/26)}$
$= 0.3154$

Sum of Many Exponential Random Variables from a Random Sample

Let $X_1, X_2, \ldots, X_n$ belong to a random sample of exponentially distributed random variables, each of which has β for the parameter.

Let $Y = X_1 + X_2 + \ldots + X_n$.

Then, Y is gamma distributed with…
    alpha = n
    beta = β

Distribution for the Time of First Occurrence among Multiple Objects with Unique Exponential Distributions

Let $X_1 \sim \text{Exp}(\beta_1)$, $X_2 \sim \text{Exp}(\beta_2)$, …, and $X_n \sim \text{Exp}(\beta_n)$.

All the X_i's are independent.

Let Y_1 = time of first occurrence among the X_i's = $\min(X_1, X_2, \ldots, X_n)$.

Then, $Y_1 \sim \text{Exp}(\beta_1 + \beta_2 + \ldots + \beta_n)$.

Example

A computer system has six processors which are working independently. The lifetime of each processor follows a unique exponential distribution. The expected lifetimes of the processors are 7, 12, 4, 6, 8, and 15 years. The computer system fails when the first chip fails.

(a.) What is the distribution for the lifetime of the computer system?
(b.) What is the expected lifetime for the system?
(c.) What's the chance the system survives more than 3 years?

Solutions:
We only have the expected lifetime of each processor. Let's invert each expected lifetime to get β_i.

$\beta_1 = 1/7$, $\beta_2 = 1/12$, $\beta_3 = 1/4$, $\beta_4 = 1/6$, $\beta_5 = 1/8$, $\beta_6 = 1/15$

(a.) Let Y_1 = lifetime of computer system
$Y_1 \sim \text{Exp}(701/840)$

(b.) Expected lifetime of system = $\dfrac{1}{701/840} = \dfrac{840}{701} = 1.1983$ years

(c.)

Digital Actuarial Resources   *Comprehensive Probability Review for Actuarial Exams*

$$\Pr(Y_1 > 3) = e^{-(701/840)*3} = 0.08179$$

Section 7-7: Beta Distribution

The beta distribution is another continuous distribution which utilizes the gamma function. A random variable with a beta distribution must have values between 0 and 1, exclusive. Since this range also corresponds to legal values of the probability for an event, the random variable is typically an unknown measure of probability. This distribution has two parameters named α and β. Both parameters must be strictly positive. When $\alpha = 1$ and $\beta = 1$, the beta distribution becomes Uniform(0, 1).

The pdf for a beta distribution contains three main terms. A leading coefficient is found by evaluating a few gamma functions. The pdf then contains two algebraic factors involving x. A common interpretation is that X is the probability of success and (1 – X) is the probability of failure. Correspondingly, α applies to the chance of success and β applies to the chance of failure. The term $x^{\alpha-1}$ is then the chance of obtaining $(\alpha-1)$ "successes," and $(1-x)^{\beta-1}$ is the chance of getting $(\beta-1)$ failures.

Probability Density Function for the Beta Distribution

$$f(x) = \begin{cases} \dfrac{\Gamma(\alpha+\beta)}{\Gamma(\alpha) \bullet \Gamma(\beta)} \bullet x^{\alpha-1} \bullet (1-x)^{\beta-1}, & \text{if } 0 < x < 1 \\ 0, & \text{else} \end{cases}$$

You can actually swap the values on α and β to obtain a beta distribution for (1 – X).

Complement of a Beta Distributed Random Variable
If X ~ Beta(α, β), then (1 – X) ~ Beta(β, α).

The graph of a beta pdf is generally a skewed bell curve. However, the plotted pdf can also resemble a line.

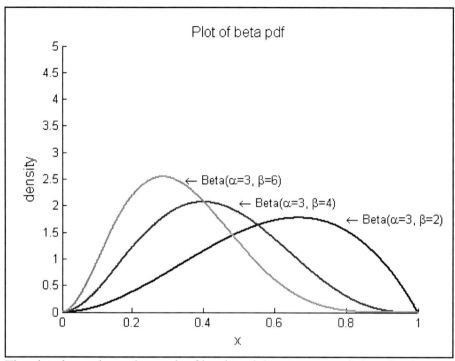

The plot above shows the result of leaving alpha constant and changing beta.

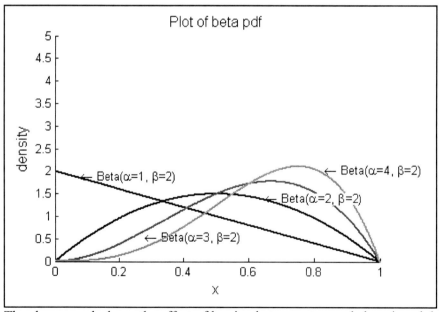

The above graph shows the effect of leaving beta constant and changing alpha.

Example
Find the pdf's for beta distributions with the following parameters:
(a.) $\alpha = 13$, $\beta = 4$ (b.) $\alpha = 5$, $\beta = 18$

Solutions:
(a.)
$$\frac{\Gamma(17)}{\Gamma(13)*\Gamma(4)} * x^{12} * (1-x)^3 = 7280 * x^{12} * (1-x)^3$$

$$f(x) = \begin{cases} 7280 * x^{12} * (1-x)^3, & \text{if } 0 < x < 1 \\ 0, & \text{else} \end{cases}$$

(b.)
$$\frac{\Gamma(23)}{\Gamma(5)*\Gamma(18)} * x^4 * (1-x)^{17} = 131{,}670 * x^4 * (1-x)^{17}$$

$$f(x) = \begin{cases} 131{,}670 * x^4 * (1-x)^{17}, & \text{if } 0 < x < 1 \\ 0, & \text{else} \end{cases}$$

Formulas for the Beta Distribution

$$E(X) = \frac{\alpha}{\alpha + \beta}$$

$$Var(X) = \frac{\alpha \beta}{(\alpha + \beta)^2 (\alpha + \beta + 1)}$$

The probability of success, p, in a binomial distribution could have a beta distribution. Let's assume the experimenter chose P to be beta distributed with initial parameters α and β (these parameters were assigned without viewing any trials). The experimenter then conducts n trials in the binomial process and observes x successes. He can revise his parameters for the distribution of P such that:

alpha = $\alpha + x$
beta = $\beta + n - x$

We now have a posterior distribution for P by using the given observations.

Example
De'Shaun is a field goal kicker in football. He does not initially know his probability of making any given kick. However, De'Shaun decides that P is beta distributed. He feels that the expected value of P is (5/6) and the variance of P is (5/684). After a few games,

he makes 7 of 11 kicks. What is his revised expected value of P after the given observations?

Solution:
First, find the parameters of the prior distribution of P without given information.

$$\frac{\alpha}{\alpha+\beta} = 5/6 \quad \rightarrow \quad \frac{\alpha+\beta}{\alpha} = 6/5 \quad \rightarrow \quad \alpha+\beta = (6/5)\alpha$$

$$\beta = (1/5)\alpha$$

$$\frac{\alpha\beta}{(\alpha+\beta)^2(\alpha+\beta+1)} = 5/684$$

$$\frac{\alpha\beta}{((6/5)\alpha)^2((6/5)\alpha+1)} = 5/684$$

$$\frac{\alpha\beta}{(6/5)^2\alpha^2((6/5)\alpha+1)} = 5/684$$

$$\frac{\beta}{(6/5)^2\alpha((6/5)\alpha+1)} = 5/684$$

$$\frac{(1/5)}{(6/5)^2((6/5)\alpha+1)} = 5/684$$

$$\frac{1}{(6/5)\alpha+1} = \frac{1}{19} \quad \rightarrow \quad (6/5)\alpha+1 = 19 \quad \rightarrow \quad \alpha = 15$$

$$\beta = (1/5)*15 = 3$$

Initially, P ~ Beta(15, 3).

In the observed binomial process, n = 11 and x = 7.

After the observations, P ~ Beta(15 + 7 = 22, 3 + 11 – 7 = 7).

Thus, the updated expected value of P is $\dfrac{22}{22+7} = 22/29$

Example
Let X equal the number of successes in a binomial process in which the chance of success (p) on any trial is unknown and the quantity of trials is 25. The value p is constant for any set of trials, but it can vary between sets. Suppose P ~ Beta(8, 2). What is the expected value of X?

Solution:

$$E(P) = \frac{8}{10}$$

$$E(X) = n * E(P) = 25 * \frac{8}{10} = 20$$

Section 7-8: Conjugate Priors

Recall the distinction between a prior and posterior distribution for a random variable. The prior distribution does not use any given information. On the other hand, the posterior distribution will utilize given observations. The observations exist in the vector $\vec{x}$. Each element is the end result of one run of the process. The observed trials are all iid. Stated differently, the observed trials form a random sample of size n. Both continuous and discrete random variables can have prior and posterior distributions. A prior pf/pdf is an initial estimate of the random variable's distribution without examining observations. A posterior pf/pdf is generally a more accurate model of the random variable's distribution after collecting observations.

This section explains how to find the posterior distribution for a random variable's parameter given a vector of observations, $\vec{x}$. Suppose you have a base random variable named X and you wish to model it with distribution D. The parameters for D can themselves be random variables. Typically, only one parameter is unknown and any other parameters are specified with constants. The unknown parameter is a random variable, which is denoted as capital Θ. A particular value of the parameter is lowercase θ. The value θ can also be said to exist in a parameter space, which is abbreviated with Ω. The prior and posterior distributions for theta and X appear as follows:

Prior/Posterior Distribution Notation

$\xi(\theta)$ represents the prior pf/pdf of theta without any given observations.

$\xi(\theta \mid \vec{x})$ is the posterior pf/pdf of theta after observing a set of outcomes embodied in $\vec{x}$.

$f(x \mid \theta)$ is the pf/pdf for a single sample point, conditional on θ.

$f_n(\vec{x} \mid \theta)$ is a joint pdf for the entire random sample, conditional on θ.

$\xi(\theta)$ is an improper prior for theta because the integral under $\xi(\theta)$ over all real values is typically not 1. In fact, the integral often evaluates to $+\infty$. The observations in $\vec{x}$ will help revise the distribution for theta, forming $\xi(\theta|\vec{x})$. The ultimate goal is to build $\xi(\theta|\vec{x})$.

The experimenter often makes an educated guess of the prior distribution for theta. Two researchers would probably produce differing estimates of the prior distribution. The observed history essentially tweaks the prior distribution, morphing it into a more accurate posterior distribution. A poor initial guess for the prior distribution usually has minimal impact on the accuracy of the posterior distribution. The level of precision in the prior distribution often fades away with a large volume of sample data.

The prior and posterior distributions for theta are related in the following likelihood equation:

Likelihood Equation

$$\xi(\theta|\vec{x}) \propto f_n(\vec{x}|\theta) * \xi(\theta)$$

The function $f_n(\vec{x}|\theta)$ is a conditional pf/pdf for X where the parameter θ is assumed to be given (but the exact value of θ might still be a mystery). The symbol $\propto$ means "is proportional to." The proportionality character is required because $f_n(\vec{x}|\theta) * \xi(\theta)$ may not represent a true pf/pdf. For example, the integral of $f_n(\vec{x}|\theta) * \xi(\theta)$ over all values of theta might not be 1. You must choose a coefficient for $f_n(\vec{x}|\theta) * \xi(\theta)$ such that a real pdf results. The proportionality character also means that you can disregard any coefficients attached to $f_n(\vec{x}|\theta)$ or $\xi(\theta)$.

The ideal way to use the likelihood equation is to try to find a template distribution with a pf/pdf similar to $\xi(\theta|\vec{x})$. A "template distribution" could be the continuous uniform distribution, geometric distribution, beta distribution, and so on. $\xi(\theta|\vec{x})$ could be identical to a template pf/pdf with the exception of an incorrect coefficient. You can simply change the coefficient attached to $\xi(\theta|\vec{x})$ in order to convert the posterior model into a true pf/pdf.

Example
Jorge runs a taco stand and is trying to model the number of customers per hour with a Poisson distribution. The expected amount of customers per hour must be 12, 16, or 20. He assigns the following prior distribution to the unknown parameter in the Poisson distribution:

$$\xi(\theta) = \begin{cases} 0.18, & \text{if } \theta = 12 \\ 0.44, & \text{if } \theta = 16 \\ 0.38, & \text{if } \theta = 20 \end{cases}$$

He runs the taco stand over the next hour and receives 19 customers. Compute the posterior distribution for the parameter in the Poisson distribution using the given observed hour.

Solution:
Let $\theta = \lambda$ = the unknown parameter
For this problem, the observed experiment has just one element, which is 19 customers received. That is, $\vec{x} = <19>$.

Evaluate the Poisson pf using x = 19 and each possible value of lambda.

$$f_n(\vec{x}|\theta = 12) = e^{-12} * \frac{12^{19}}{19!} = 0.01614$$

$$f_n(\vec{x}|\theta = 16) = e^{-16} * \frac{16^{19}}{19!} = 0.0699$$

$$f_n(\vec{x}|\theta = 20) = e^{-20} * \frac{20^{19}}{19!} = 0.08884$$

$$\xi(\theta = 12 | \vec{x}) \propto f_n(\vec{x}|\theta = 12) * \xi(\theta = 12) = 0.01614 * 0.18 = 0.00291$$
$$\xi(\theta = 16 | \vec{x}) \propto f_n(\vec{x}|\theta = 16) * \xi(\theta = 16) = 0.0699 * 0.44 = 0.03076$$
$$\xi(\theta = 20 | \vec{x}) \propto f_n(\vec{x}|\theta = 20) * \xi(\theta = 20) = 0.08884 * 0.38 = 0.03376$$

The posterior pf does not yet sum to 1. We must normalize the probabilities.

$$\xi(\theta = 12 | \vec{x}) = 0.04316$$
$$\xi(\theta = 16 | \vec{x}) = 0.45618$$
$$\xi(\theta = 20 | \vec{x}) = 0.50067$$

Example
Dr. Greenbaum is a psychologist prescribing lithium to bipolar patients. He believes that the chance of the drug successfully working for any given patient is 0.6, 0.75, or 0.9, all with the same probability initially. He gives the drug to 16 patients and finds that the drug worked with 13 of the patients. What should Dr. Greenbaum use for his updated estimates of the chances the drug works with any patient? Assume that all patients are identical and independent.

Solution:
Let θ = probability the drug cures any given patient
The prior pf of θ is:

$$\xi(\theta) = \begin{cases} 1/3, & \text{if } \theta = 0.6 \\ 1/3, & \text{if } \theta = 0.75 \\ 1/3, & \text{if } \theta = 0.9 \end{cases}$$

The observed random sample has size 16. Additionally, 13 successes occur. Since the success/failure of each patient can be modeled as a Bernoulli random variable, and all the patients are independent, the random sample is equivalent to a binomial process. We are concerned about the number of successes out of 16.

$$f_n(\vec{x}|\theta = 0.6) = \binom{16}{13} * 0.6^{13} * 0.4^3 = 0.04681$$

$$f_n(\vec{x}|\theta = 0.75) = \binom{16}{13} * 0.75^{13} * 0.25^3 = 0.20788$$

$$f_n(\vec{x}|\theta = 0.9) = \binom{16}{13} * 0.9^{13} * 0.1^3 = 0.14234$$

$\xi(\theta = 0.6 | \vec{x}) \propto f_n(\vec{x}|\theta = 0.6) * \xi(\theta = 0.6) = 0.04681 * (1/3) = 0.0156$
$\xi(\theta = 0.75 | \vec{x}) \propto f_n(\vec{x}|\theta = 0.75) * \xi(\theta = 0.75) = 0.20788 * (1/3) = 0.06929$
$\xi(\theta = 0.9 | \vec{x}) \propto f_n(\vec{x}|\theta = 0.9) * \xi(\theta = 0.9) = 0.14234 * (1/3) = 0.04745$

$\xi(\theta | \vec{x})$ is not quite a pf yet because the probabilities do not sum to 1. They need to be normalized. You can normalize each posterior probability by dividing by the sum of all the probabilities.

$$\xi(\theta | \vec{x}) = \begin{cases} 0.11788, & \text{if } \theta = 0.6 \\ 0.52358, & \text{if } \theta = 0.75 \\ 0.35855, & \text{if } \theta = 0.9 \end{cases}$$

The experimental trials by Dr. Greenbaum yielded a success rate of 81.25%. We should expect the value of theta closest to 81.25% to receive the largest bump up in posterior probability. In addition, the value of theta farthest from 81.25% should get the greatest drop in posterior probability. Correspondingly, $\theta = 0.75$ now has a probability value over 50%, and $\theta = 0.6$ endured a probability drop down to about 12%.

Example

Dredge built a computer security system meant to keep out intruders. He believes the chance the system will stop any given intrusion attempt is 95%, 96%, or 99%. Dredge built the following prior distribution for the unknown chance, θ, that the system can stop an intrusion:

$$\xi(\theta) = \begin{cases} 0.25, & \text{if } \theta = 0.95 \\ 0.60, & \text{if } \theta = 0.96 \\ 0.15, & \text{if } \theta = 0.99 \end{cases}$$

He then tests the system and finds that it stopped 34 of 35 attempted attacks. What should Dredge use for his posterior distribution of θ?

Solution:
The observed trials form a random sample of size 35. The number of successes in the random sample has a binomial distribution.

$$f_n(\vec{x} | \theta = 0.95) = \binom{35}{34} * 0.95^{34} * 0.05^1 = 0.30594$$

$$f_n(\vec{x} | \theta = 0.96) = \binom{35}{34} * 0.96^{34} * 0.04^1 = 0.34942$$

$$f_n(\vec{x} | \theta = 0.99) = \binom{35}{34} * 0.99^{34} * 0.01^1 = 0.24869$$

$\xi(\theta = 0.95 | \vec{x}) \propto f_n(\vec{x} | \theta = 0.95) * \xi(\theta = 0.95) = 0.30594 * 0.25 = 0.07649$
$\xi(\theta = 0.96 | \vec{x}) \propto f_n(\vec{x} | \theta = 0.96) * \xi(\theta = 0.96) = 0.34942 * 0.60 = 0.20965$
$\xi(\theta = 0.99 | \vec{x}) \propto f_n(\vec{x} | \theta = 0.99) * \xi(\theta = 0.99) = 0.24869 * 0.15 = 0.0373$

The posterior pf of theta is not yet normalized.

$$\xi(\theta | \vec{x}) = \begin{cases} 0.23649, & \text{if } \theta = 0.95 \\ 0.64819, & \text{if } \theta = 0.96 \\ 0.11532, & \text{if } \theta = 0.99 \end{cases}$$

Example
The random variable X is uniformly distributed on [100, θ]. The upper bound is a random variable θ, with a prior distribution obeying Uniform(140, 160). The experimenter observes one value of X, which is x = 155. Find θ's posterior distribution.

Solution:
The sample has just one observation, which is x = 155.

$$f_n(\vec{x}|\theta) = f(\vec{x}|\theta) = \frac{1}{\theta - 100} = \text{a real value}$$

$$\xi(\theta) = \frac{1}{160 - 140} = \frac{1}{20}$$

$$\xi(\theta|\vec{x}) \propto \frac{1}{\theta - 100} * \frac{1}{20} \propto 1$$

The posterior distribution for theta must be a continuous uniform distribution. The bounds on the distribution are still unknown.

At this point, we can find the posterior bounds on theta using logic. The observed value was 155. If theta was below 155, then the value x = 155 could never be obtained. For example, if X must be uniformly distributed on [100, 150], then x = 155 would be impossible. Therefore, theta must exceed or equal 155. x = 155 would only be possible if theta belongs to [155, 160].

Therefore,
$$\xi(\theta|\vec{x}) = \begin{cases} 1/5, & \text{if } 155 \leq \theta \leq 160 \\ 0, & \text{else} \end{cases}$$

If you can carefully choose the prior distribution for Θ, then determining a posterior distribution for the parameter can be quite easy. The prior and posterior distributions could belong to the same general family. Shortcuts exist when the chosen prior distribution for Θ fits in a "conjugate family" for the base random variable X.

Conjugate Priors for the Normal Distribution

Let the base random variable, X, have a normal distribution. We know the variance of X, which is σ^2, but the expected value of X is unknown. Let Θ denote the random variable representing the mean of X. A convenient choice for theta's distribution is a normal distribution. Let θ have a prior, normal distribution with expected value μ_1

and variance v_1^2. Then, the posterior distribution of θ is also normal with the following parameters:

$$\mu_2 = \frac{\sigma^2}{\sigma^2 + nv_1^2}\mu_1 + \frac{nv_1^2}{\sigma^2 + nv_1^2}\overline{x}_n \qquad v_2^2 = \frac{\sigma^2 v_1^2}{\sigma^2 + nv_1^2}$$

Notice that the expected value of the posterior distribution is a weighted average of means. The two means being weighted are μ_1 and $\overline{x}_n$. The parameters μ_1 and v_1^2 are termed "prior hyperparameters." Correspondingly, the parameters μ_2 and v_2^2 are called "posterior hyperparameters." The "hyperparameter" terms apply to any situation (not just normal distributions) in which the parameter for the base random variable itself has parameters.

Example
The random variable X is normally distributed. The mean of X, denoted θ, is still a mystery, but the variance of X is 49. The experimenter gathers a random sample with 42 observations. The sample mean is 28. The experimenter believes the prior distribution of θ is normal with parameters $\mu_1 = 25$ and $v_1^2 = 36$. What is the posterior distribution of θ?

Solution:
$\sigma^2 = 49$, $n = 42$, $\overline{x}_n = 28$, $\mu_1 = 25$, $v_1^2 = 36$

The posterior distribution of θ is normal with parameters

$$\mu_2 = \frac{49}{49 + 42*36}*25 + \frac{42*36}{49 + 42*36}*28 = 27.9058$$

$$v_2^2 = \frac{49*36}{49 + 42*36} = 1.13$$

Example
X has a normal distribution with an unspecified mean, called θ, and a variance of 81. The experimenter collects a random sample of size 6 and finds the sample mean to be 150. The experimenter later discovers that the posterior distribution of θ is normal with mean 145 and variance 4. Compute the prior distribution of θ.

Solution:
$\sigma^2 = 81$, $n = 6$, $\overline{x}_n = 150$, $\mu_2 = 145$, $v_2^2 = 4$

Since the posterior distribution of θ is normal, the prior distribution of θ must also have been normal. The parameters for the prior distribution of θ are solved below:

$$4 = \frac{81 * v_1^2}{81 + 6 * v_1^2} \quad \rightarrow \quad 324 + 24v_1^2 = 81 * v_1^2 \quad \rightarrow \quad v_1^2 = 5.6842$$

$$145 = \frac{81}{81 + 6 * 5.6842} \mu_1 + \frac{6 * 5.6842}{81 + 6 * 5.6842} * 150 \quad \rightarrow \quad \mu_1 = 142.8947$$

Conjugate Priors for the Exponential Distribution

Let X be a random variable with an exponential distribution. The value of the parameter β within the exponential distribution is an unidentified random variable. A sound choice for the prior distribution of β is a gamma distribution with parameters α_1 and β_1. After we observe a sequence of X_i observations, we can model the posterior distribution of β with another gamma distribution. The posterior gamma distribution will have parameters

$$\alpha_2 = \alpha_1 + n \qquad \beta_2 = \beta_1 + \sum x_i$$

Overall, the gamma distribution forms a conjugate family for the exponential distribution.

Example
The random variable X has an exponential distribution with unknown parameter β. The experimenter decides to model the prior distribution of β with a gamma distribution having parameters $\alpha_1 = 17$ and $\beta_1 = 187$. He gathers the following vector of observations:

$$\vec{x} = <9, 14, 11, 12, 10, 9>$$

What is the posterior distribution of β?

Solution:
The posterior distribution of β is gamma with these parameters:
$$\alpha_2 = 17 + 6 = 23$$
$$\beta_2 = 187 + 65 = 252$$

Conjugate Priors for the Bernoulli Distribution

Assume that the base random variable of interest has a Bernoulli distribution, and the value of the parameter p is unknown. The beta distribution forms a conjugate family for the Bernoulli distribution. That is, we will assume p has a prior distribution which is beta with specified parameters α_1 and β_1. After observing a series of n independent trials, each of which could succeed or fail, we can compute the posterior distribution. The posterior distribution will also be beta, but the parameters are set as follows:

$$\alpha_2 = \alpha_1 + (\text{\# successes})$$
$$\beta_2 = \beta_1 + (\text{\# failures})$$

To form the parameters for the posterior beta distribution, the original parameters from the prior distribution are morphed depending on how many successes and failures happened. Alpha corresponds to the relative number of successes, while beta corresponds to the relative amount of failures.

The beta distribution is an excellent choice for the prior and posterior distributions of the parameter p in a Bernoulli distribution because p must be between 0 and 1, and the random variable in a beta distribution is restricted between 0 and 1.

Example
Jasmine is shooting hoops with a basketball. She does not know her chance of success in making any given shot. She believes that the chance of making a shot has a beta distribution with parameters $\alpha_1 = 7$ and $\beta_1 = 2$. She throws the ball 24 independent times and makes 8 baskets.

(a.) What is the posterior distribution for p given the observed shots?
(b.) Compute the prior and posterior expected values of p.

Solutions:
(a.) n = 24 shots, (# successes) = 8, (# failures) = 16

$\xi(p \mid \vec{x})$ represents the pdf for a beta distribution with parameters

$$\alpha_2 = 7 + 8 = 15 \quad \text{and} \quad \beta_2 = 2 + 16 = 18$$

(b.)
Prior expected value of $p = \dfrac{7}{7+2} = 7/9$

Posterior expected value of p = $\dfrac{15}{15+18} = 15/33 = 5/11$

Conjugate Priors for the Poisson Distribution

Let X have a Poisson distribution. The parameter λ is unknown, and we desire to estimate a distribution for lambda. The gamma distribution forms a conjugate family for the Poisson distribution. A good choice for the prior distribution of λ is a gamma distribution with parameters α_1 and β_1. The posterior distribution of λ will also be gamma with the following parameters:

$$\alpha_2 = \alpha_1 + \sum x_i$$
$$\beta_2 = \beta_1 + n$$

Each outcome, x_i, stores the number of occurrences observed in one time interval of the Poisson process. Once again, there are n total sample points.

Example
Security at a high-tech company is always on the watch for intruders. Officers believe that the number of intrusions during a night has a Poisson distribution, but they don't know the expected amount of intrusions per night. Without utilizing data, the officers initially attempt to model lambda with a gamma distribution with parameters $\alpha_1 = 48$ and $\beta_1 = 4$. They decide to monitor the building every night for a week. They observe the following numbers of intrusions (each element is the quantity of intrusions for a particular night):

$$\vec{x} = <9, 14, 10, 12, 16, 5, 11>$$

(a.) Determine the posterior distribution for the parameter λ.
(b.) What are the prior and posterior expected values of λ?

Solutions:
(a.)
$n = 7$, $\sum x_i = 77$

The posterior distribution of lambda is gamma with…

$$\alpha_2 = 48 + 77 = 125 \quad \text{and} \quad \beta_2 = 4 + 7 = 11$$

(b.)

> prior expected value of $\lambda = \dfrac{\alpha_1}{\beta_1} = 12$
>
> posterior expected value of $\lambda = \dfrac{\alpha_2}{\beta_2} = 11.3636$

The table below summarizes the conjugate families:

Table of Conjugate Priors

base distribution for X	definition of θ	prior distribution of θ	posterior distribution of θ
Normal	mean	Normal(μ_1, v_1^2)	Normal with $\mu_2 = \dfrac{\sigma^2}{\sigma^2 + nv_1^2}\mu_1 + \dfrac{nv_1^2}{\sigma^2 + nv_1^2}\bar{x}_n$ $v_2^2 = \dfrac{\sigma^2 v_1^2}{\sigma^2 + nv_1^2}$
Exponential	β	Gamma(α_1, β_1)	Gamma with $\alpha_2 = \alpha_1 + n$ $\beta_2 = \beta_1 + \sum x_i$
Bernoulli	p	Beta(α_1, β_1)	Beta with $\alpha_2 = \alpha_1 + $ (# successes) $\beta_2 = \beta_1 + $ (# failures)
Poisson	λ	Gamma(α_1, β_1)	Gamma with $\alpha_2 = \alpha_1 + \sum x_i$ $\beta_2 = \beta_1 + n$

Appendix A: Standard Normal Distribution

z	F(z)	z	F(z)	z	F(z)	z	F(z)	z	F(z)	z	F(z)	z	F(z)
0.00	0.5000	0.50	0.6915	1.00	0.8413	1.50	0.9332	2.00	0.9772	2.50	0.9938	3.00	0.9987
0.01	0.5040	0.51	0.6950	1.01	0.8438	1.51	0.9345	2.01	0.9778	2.51	0.9940	3.01	0.9987
0.02	0.5080	0.52	0.6985	1.02	0.8461	1.52	0.9357	2.02	0.9783	2.52	0.9941	3.02	0.9987
0.03	0.5120	0.53	0.7019	1.03	0.8485	1.53	0.9370	2.03	0.9788	2.53	0.9943	3.03	0.9988
0.04	0.5160	0.54	0.7054	1.04	0.8508	1.54	0.9382	2.04	0.9793	2.54	0.9945	3.04	0.9988
0.05	0.5199	0.55	0.7088	1.05	0.8531	1.55	0.9394	2.05	0.9798	2.55	0.9946	3.05	0.9989
0.06	0.5239	0.56	0.7123	1.06	0.8554	1.56	0.9406	2.06	0.9803	2.56	0.9948	3.06	0.9989
0.07	0.5279	0.57	0.7157	1.07	0.8577	1.57	0.9418	2.07	0.9808	2.57	0.9949	3.07	0.9989
0.08	0.5319	0.58	0.7190	1.08	0.8599	1.58	0.9429	2.08	0.9812	2.58	0.9951	3.08	0.9990
0.09	0.5359	0.59	0.7224	1.09	0.8621	1.59	0.9441	2.09	0.9817	2.59	0.9952	3.09	0.9990
0.10	0.5398	0.60	0.7257	1.10	0.8643	1.60	0.9452	2.10	0.9821	2.60	0.9953	3.10	0.9990
0.11	0.5438	0.61	0.7291	1.11	0.8665	1.61	0.9463	2.11	0.9826	2.61	0.9955	3.11	0.9991
0.12	0.5478	0.62	0.7324	1.12	0.8686	1.62	0.9474	2.12	0.9830	2.62	0.9956	3.12	0.9991
0.13	0.5517	0.63	0.7357	1.13	0.8708	1.63	0.9484	2.13	0.9834	2.63	0.9957	3.13	0.9991
0.14	0.5557	0.64	0.7389	1.14	0.8729	1.64	0.9495	2.14	0.9838	2.64	0.9959	3.14	0.9992
0.15	0.5596	0.65	0.7422	1.15	0.8749	1.65	0.9505	2.15	0.9842	2.65	0.9960	3.15	0.9992
0.16	0.5636	0.66	0.7454	1.16	0.8770	1.66	0.9515	2.16	0.9846	2.66	0.9961	3.16	0.9992
0.17	0.5675	0.67	0.7486	1.17	0.8790	1.67	0.9525	2.17	0.9850	2.67	0.9962	3.17	0.9992
0.18	0.5714	0.68	0.7517	1.18	0.8810	1.68	0.9535	2.18	0.9854	2.68	0.9963	3.18	0.9993
0.19	0.5753	0.69	0.7549	1.19	0.8830	1.69	0.9545	2.19	0.9857	2.69	0.9964	3.19	0.9993
0.20	0.5793	0.70	0.7580	1.20	0.8849	1.70	0.9554	2.20	0.9861	2.70	0.9965	3.20	0.9993
0.21	0.5832	0.71	0.7611	1.21	0.8869	1.71	0.9564	2.21	0.9864	2.71	0.9966	3.21	0.9993
0.22	0.5871	0.72	0.7642	1.22	0.8888	1.72	0.9573	2.22	0.9868	2.72	0.9967	3.22	0.9994
0.23	0.5910	0.73	0.7673	1.23	0.8907	1.73	0.9582	2.23	0.9871	2.73	0.9968	3.23	0.9994
0.24	0.5948	0.74	0.7704	1.24	0.8925	1.74	0.9591	2.24	0.9875	2.74	0.9969	3.24	0.9994
0.25	0.5987	0.75	0.7734	1.25	0.8944	1.75	0.9599	2.25	0.9878	2.75	0.9970	3.25	0.9994
0.26	0.6026	0.76	0.7764	1.26	0.8962	1.76	0.9608	2.26	0.9881	2.76	0.9971	3.26	0.9994
0.27	0.6064	0.77	0.7794	1.27	0.8980	1.77	0.9616	2.27	0.9884	2.77	0.9972	3.27	0.9995
0.28	0.6103	0.78	0.7823	1.28	0.8997	1.78	0.9625	2.28	0.9887	2.78	0.9973	3.28	0.9995
0.29	0.6141	0.79	0.7852	1.29	0.9015	1.79	0.9633	2.29	0.9890	2.79	0.9974	3.29	0.9995
0.30	0.6179	0.80	0.7881	1.30	0.9032	1.80	0.9641	2.30	0.9893	2.80	0.9974	3.30	0.9995
0.31	0.6217	0.81	0.7910	1.31	0.9049	1.81	0.9649	2.31	0.9896	2.81	0.9975	3.31	0.9995
0.32	0.6255	0.82	0.7939	1.32	0.9066	1.82	0.9656	2.32	0.9898	2.82	0.9976	3.32	0.9995
0.33	0.6293	0.83	0.7967	1.33	0.9082	1.83	0.9664	2.33	0.9901	2.83	0.9977	3.33	0.9996
0.34	0.6331	0.84	0.7995	1.34	0.9099	1.84	0.9671	2.34	0.9904	2.84	0.9977	3.34	0.9996
0.35	0.6368	0.85	0.8023	1.35	0.9115	1.85	0.9678	2.35	0.9906	2.85	0.9978	3.35	0.9996
0.36	0.6406	0.86	0.8051	1.36	0.9131	1.86	0.9686	2.36	0.9909	2.86	0.9979	3.36	0.9996
0.37	0.6443	0.87	0.8079	1.37	0.9147	1.87	0.9693	2.37	0.9911	2.87	0.9979	3.37	0.9996
0.38	0.6480	0.88	0.8106	1.38	0.9162	1.88	0.9699	2.38	0.9913	2.88	0.9980	3.38	0.9996
0.39	0.6517	0.89	0.8133	1.39	0.9177	1.89	0.9706	2.39	0.9916	2.89	0.9981	3.39	0.9997
0.40	0.6554	0.90	0.8159	1.40	0.9192	1.90	0.9713	2.40	0.9918	2.90	0.9981	3.40	0.9997
0.41	0.6591	0.91	0.8186	1.41	0.9207	1.91	0.9719	2.41	0.9920	2.91	0.9982	3.41	0.9997
0.42	0.6628	0.92	0.8212	1.42	0.9222	1.92	0.9726	2.42	0.9922	2.92	0.9982	3.42	0.9997
0.43	0.6664	0.93	0.8238	1.43	0.9236	1.93	0.9732	2.43	0.9925	2.93	0.9983	3.43	0.9997
0.44	0.6700	0.94	0.8264	1.44	0.9251	1.94	0.9738	2.44	0.9927	2.94	0.9984	3.44	0.9997
0.45	0.6736	0.95	0.8289	1.45	0.9265	1.95	0.9744	2.45	0.9929	2.95	0.9984	3.45	0.9997
0.46	0.6772	0.96	0.8315	1.46	0.9279	1.96	0.9750	2.46	0.9931	2.96	0.9985	3.46	0.9997
0.47	0.6808	0.97	0.8340	1.47	0.9292	1.97	0.9756	2.47	0.9932	2.97	0.9985	3.47	0.9997
0.48	0.6844	0.98	0.8365	1.48	0.9306	1.98	0.9761	2.48	0.9934	2.98	0.9986	3.48	0.9997
0.49	0.6879	0.99	0.8389	1.49	0.9319	1.99	0.9767	2.49	0.9936	2.99	0.9986	3.49	0.9998

Appendix B: Common Series

Common Arithmetic Series

$$\sum_{i=0}^{n} i = \frac{n(n+1)}{2}$$

$$\sum_{i=0}^{n} i^2 = \frac{n(n+1)(2n+1)}{6}$$

$$\sum_{i=0}^{n} i^3 = \frac{n^2(n+1)^2}{4}$$

Common Geometric Series
Let r represent a rate of change between terms.

$$\sum_{i=0}^{n} r^i = \frac{r^{n+1}-1}{r-1}$$

$$\sum_{i=0}^{\infty} r^i = \frac{1}{1-r}, \text{ assuming } -1 < r < 1$$

CPSIA information can be obtained at www.ICGtesting.com
Printed in the USA
BVOW052142130812

297799BV00001B/1/P

9 781453 780794